SO-AFS-183

SNOW CAVES
FOR
FUN AND SURVIVAL

by
Ernest Wilkinson

COPYRIGHTED 1986
ISBN NUMBER
0-912510 - 03-X

No part of this book may be reproduced in any
manner whatsoever without written permission
from the publisher

First Printing 1986

PUBLISHED BY WINDSONG PRESS
P. O. Box 1484
Denver, CO 80201

Printed by Jostens Printing & Publishing Division
TOPEKA, KS

ACKNOWLEDGEMENTS

Writing this book has been a new experience for me and its completion is due to the efforts of several special people. I especially want to thank my wife, Margaret, and my son, Larry, for the many hours they spent typing and correcting the spelling and punctuation of my rough manuscript plus their patience with me during my many nights of experimenting with snowcaves.

I also wish to thank the publisher, Hal Webster, for his patience and time of working with me to polish my "country boy" writing after he learned of my experience with the subject. Credit should also go to the many participants of my winter tours that furnished me the opportunity to observe, study, and test the actions of assorted winter campers.

BIOGRAPHY

The author has spent many years living and traveling alone in the mountains of Colorado. During the summer months he worked as a government trapper controlling bear and coyotes that were killing livestock in the high summer ranges and conducted tours in the mountains during the winter plus some high country trapping.

He and his wife, Margaret, raised and trained mountain lions, coyotes, badgers, and other animals for filming educational and TV programs. (Wild Kingdom, Wild Wild World of Animals, Encyclopaedia Britannica, and others). They learned a lot about survival and improvising from nature by observing the various species of wildlife. He was coordinator for the San Luis Valley Search & Rescue Association for numerous years and during the many missions realized there was a great need for wilderness survival education by the general public. He began to give lectures and programs on survival to schools and other groups.

Not satisfied with the methods of snowcave construction shown in most of the winter survival manuals, he developed his own time and energy saving technic for snowcaves as body protection and added comfort to winter outdoor excursions. These methods were tested and proven satisfactory with hundreds of students on his winter tours. He felt the general public needed more knowledge or training on how to better take care of themselves during winter conditions in emergency situations or during pleasure or work excursions, hence the reason for this book in the hopes of assisting outdoor users.

Along with operating a taxidermy shop west of Monte Vista, Colorado and filming wildlife for educational uses, he still continues to conduct winter tours showing the use of snowcaves to assorted groups.

CONTENTS

SNOW SHELTERS
FOR
FUN AND SURVIVAL

by
Ernest Wilkinson

CHAPTER 1
WHY?

How many rabbits have you seen sitting by a fire to keep warm? This may sound like a silly question, but give it some thought and you will begin to understand the principal involved. The human critter, man, homo-sapien, or whatever name you prefer to use has become dependent upon a fire source for body comfort in a cold environment. Hunters, hikers, and other outdoor users have perished because they could not get a fire started to help keep them warm in cold weather. They should have taken a lesson from rabbits and other wildlife that have learned to utilize the available materials of nature. During storms or cold weather, wild animals and birds like the ptarmigan often utilize a small hole in the ground or snow to help retain body heat. They curl up in a small burrow and their own body heat soon warms the area. Dry grass or leaves are often utilized for warmth.

Not so with man! Most of us have become accustomed to large, spacious living quarters where, if not protected by other means, our body heat radiates out to the atmosphere and is lost. During an unforeseen situation in a cold environment, this can be fatal if a fire or other heat producing source is not available. During the winter in high country areas, snow is usually available and is one of nature's better insulating materials. So why not utilize it for camping comfort or when there is a need to conserve body heat instead of letting that warmth dissipate into the atmosphere?

When you wear a garment made of down or synthetic fibers, it is the many small pockets of air that form an insulation barrier around you. Snow is generally composed of sixty to ninety percent or more of air between the flakes as they fall. The air content of this snow is usually much higher during the first part of the winter before the sun, wind, and weight of more snow gradually compact the lower layers. No matter what time of the winter or condition of the snow, it can be utilized for snowcaves, igloos, and other improvised body shelters because of the tiny air pockets in the snow.

Before describing these snow shelters, it is best that we first understand the basic principles of how the human body functions and how it maintains itself in a cold environment. The physical and mental condition of the individual are both important. Each winter more skiers, snowshoers, snowmobilers, and other recreationists venture into remote areas where they are not properly prepared to do so.

More and more people are participating in winter excursions into remote areas.

Because of "push button" living where one can flip a switch for lights, turn a faucet for water, or turn a thermostat knob for heat, many individuals are unable to improvise for themselves during an emergency situation when those switches or controls are not available. The general public has not been conditioned to prepare for an extreme situation for many reasons. We live and work in a basically controlled environment. We live in insulated shelters with readily available heat, water and lights. We walk a few feet to cars or buses that have heaters. These cars or buses travel on paved roads to places of business or employment which are likely to be a regulated condition. Our clothes are designed for creating a good public appearance rather than for practical use in an outdoor situation should the need arise.

The urban life style does not prepare the average individual for good physical or mental health. Many of us are creatures of habit in our day to day lifestyle. We arise at a certain time in the morning, eat a hurried breakfast (if any at all), and rush through congested traffic to arrive at an office or place of employment which usually has a maintained temperature. We ride elevators or esculators instead of climbing stairs, take buses or drive cars instead of walking or riding bikes. These things certainly don't enhance our physical condition.

I personally believe we can receive better intellectual stimulation and cultivation when our bodies have been exercised rather than sitting behind a desk all day, utilizing elevators and vehicles to go from place to place, and then sitting in front of a television all evening.

In addition to physical conditioning we need to know what it takes to keep our bodies operating efficiently in a cold environment and how to plan ahead to cope with emergency situations. The food, water, and air that we take into our systems goes through various processes to emerge as body energy or body heat. Our stomach is similar to the fuel tank in a car. There is a certain amount of fuel there and it can take you far or near depending on how you drive.

Planning ahead is important. If you use up all of this fuel as energy to get to the top of the hill or to struggle through deep snow, there might not be any fuel left for body heat if you have to spend the night out under a tree or in a make-shift shelter. You can go much further at a slow, steady pace than rushing around and ending up totally exhausted. It is mandatory that you take a few minutes and make a plan. This plan should include not only your travel direction and speed but also a shelter. This shelter will consist of the clothing you are wearing, items you have with you, and improvised materials that mother nature will provide.

Before describing the construction of snow shelters, it is important to explain hypothermia so you can better understand why

9

adequate shelters are needed. Hypothermia is a condition when your body core temperature is subnormal because it is losing heat faster than it is being created. The normal body temperature is 98.6 degrees which can vary slighthly due to age or metabolism. Body heat can be rapidly lost when one is exposed to a cold wind and wet or improper clothing. If you find yourself starting to shiver, fingers becoming numb and you have difficulty in picking up a small object or attempting to fasten a button, this is an indication that you are losing body heat and hypothermia could already be starting.

When shivering starts and your fingers begin to numb, your body core temperature is about 97 degrees. Shivering is an automatic body response making an effort to create body heat. Rubbing causes friction and friction causes heat. Your muscles contract or twitch to create heat for your body but shivering unfortunately uses more energy than normal. At this stage, your brain has not yet been affected by the cold. So heed those body indicators and take immediate actions to stop that loss of body temperature by getting protection from the wind by moving off of a ridge or taking shelter behind rocks, brush or other natural windbreaks. You can also use more dry clothing, more fuel into the stomach, and even some rest in a protected area to let the body recuperate.

When your body core temperature begins to lower, the flow of blood to the extremeties is restricted and shunted to the vital organs (heart, lungs, liver and kidneys) to keep them warm. This is part of the reason your toes and fingers can freeze so quickly. Your body takes compensatory measures because you can get by without a foot or finger, but life is short without a heart.

When wet some types of clothing can lose up to ninety percent of their insulation value, so changing into dry clothing can help to stop the lose of body heat. Large amounts of body heat can be lost through an uncovered head. Make sure your head and neck are properly protected. Under some conditions with wet clothing and a cold wind blowing, you can lose body heat 24 times faster than normal. Attempt to stay dry and reduce exposure to the wind. If you do not stop this drop in body-core temperature, the flow of blood to the extremities will be increasingly restrained so that cold blood will not be carried back to the vital organs. If this downward trend of body core temperature continues, the flow of blood to the head is restricted, decreasing the supply of oxygen to the brain. The brain uses approximately 25 percent of the entire body oxygen supply, consequently, your powers to think decreases and you are slowly losing the ability to help yourself.

This process comes on so gradually that the victim does not realize it is happening. This is the reason you should take corrective steps to stop the loss of body heat when your fingers become numb or you start to shiver. You still have a brain to think with .... use it! You will feel these body indicators before your buddies can visually detect them. So, be alert! When the flow of blood to the brain is restricted, the victims speech becomes slurry and they develop an "I don't give a darn attitude". They become careless about keeping their head cover or gloves on. He or she begins to stumble more often. At this stage the victim is beyond self-help so the "buddy system" is very important. When a companion notices any of the hypothermia indicators in another person they should immediately take steps to get the victim into dry clothing, out of the wind, and other procedures such as getting warm liquids into the system to stop that downward trend of body core temperature.

After hypothermia has dulled the brain, you can be freezing with a sleeping bag or dry garments near by and you do not have brain function enough to use them. At this stage, when your brain becomes numb, your body core temperature is about 95 degrees. By the time the body core temperature drops to 92 degrees, the ability for self-help is completely gone with unconsciousness coming at approximately 87 degrees. This bottom line temperature can vary slightly with individuals.

You can now realize that the human body has a very narrow range of body temperature efficiency based on a normal temperature of 98.6 degrees. You might find it difficult to understand what losing a few degrees of body temperature can do and you might say, "No, I could never be dumb enough to not use a sleeping bag in my pack or a jacket laying beside me". But believe me, it does happen! I could not visualize a person freezing and not using the available items. After actually seeing hypothermia victims during local search and rescue missions, I better understand and respect hypothermia.

As an example, our search and rescue crew received a request from a local sheriff concerning a lost hunter in the Goose Creek area in Colorado. When finally found in the snow-covered mountains, the man was dead in a face-down position on the creek bank. His shoes, hat, and jacket were gone. We can only imagine what actually happened. Perhaps when he became disoriented and lost from camp during the snow storm, panic must have taken over as he rushed back and forth from one ridge to another in an attempt to locate a recognizable landmark. His underclothes probably became damp with perspiration and he might have even gotten his feet wet when crossing the creek or by wading through the falling snow. With the cold wind and his underclothes damp with perspiration, his remaining body heat was being needlessly lost into the atmosphere. It appeared that he had taken his shoes off and probably rubbed his feet to warm them. During this process, hypothermia crept further into his system thus dulling the brain. With the brain dulled, he apparently wondered off through the falling snow, leaving his shoes behind. As his jacket and hat became wet and cumbersome this burden was also discarded along the wayside. Without these items to protect against body heat loss, unconsciousness soon set in and he ended up where we found him.

Not knowing the damaging processes of hypothermia, he had waited too long to take any corrective action. Finally he did not have enough brain power left to even keep his shoes and jacket on to help retain body heat.

On another occasion during midwinter, as coordinator for San Luis Valley Search and Rescue Association, I received a phone call one evening from the Rio Grande County Sheriff stating that a local sheepherder was long overdue. He could not be located by the sheep owner whom had left him at 2:30 that afternoon coming down the mountain toward the ranch with 540 sheep. A blizzard had moved in late that afternoon and neither the man nor the sheep could be located as darkness fell.

I gathered together a crew of six expert snowmobilers and we met with the sheriff and the sheep owner at 8:00 P.M. for a search briefing. We drove toward the foothill area until snowdrifts blocked the road. The snowmobiles were unloaded and we fanned out in pairs into the swirling snow. The snowmobile lights probed a short distance ahead into the blackness as we steered the machines through the sage brush and across numerous gullies. We finally located the sheep bedded down out the of the wind in a low spot, but we could not locate the herder. Several sets of what appeared to be man tracks in the drifting snow were located, but they would soon be obliterated by the blowing snow when they crossed any open area.

To make a long story short, the wind quit blowing at 10:30 P.M. and it quit snowing at 11:00 P.M. with the stars starting to shine brightly. Without the cloud cover to retain heat, the temperature dropped rapidly to 32 degrees below zero. While making a wide circle in the darkness, we came across a set of man tracks that we could easily follow as the wind was not blowing the tracks out. One of the searchers got within ten feet of the walking, but occasionally stumbling, sheepherder and called out his name. No response at all, so he took the disoriented man by the shoulders, turned him down-hill and guided him to where we finally got him into the sheriff's car and straight away to the hospital.

This sheepherder had lived in that area most of his life and knew the terrain very well. When it finally quit snowing, he could see the various ranch lights below him across the foothills, but he made no attempt to go towards them. With his brain power diminished, he just continued to wonder aimlessly across the rolling hills totally disregarding the ranch lights. If the search and rescue crew had not located him when they did, the man could not have lasted much longer. After being warmed at the hospital, the man bounced back to normal except for some partially frosted fingers and toes.

When traveling or working in any dampness or chilly weather, remember to take care of the brain and the brain will take care of the body. If you wait too long to correct that drop in body temperature as hypothermia progresses you might then be unable to construct a body shelter from available material for protection. Hypothermia is known as the silent killer and is responsible for numerous outdoor deaths, especially in cold or damp weather conditions. Knowing the symptoms of hypothermia and how to correct the drop in body core temperature can be very important in an emergency situation.

In the above mentioned incidents neither of the victims had partners along to assist them when their thinking power diminished. This illustrates why the buddy system is very important and is another reason why you should not venture alone and unprepared into the remote areas.

When assisting a hypothermia victim, a tent from your pack can give additional shelter from the wind. If hypothermia has progressed and the victim is in trouble, skin to skin contact is very helpful in getting a body warmed up and functioning normally again. If a sleeping bag is available, get the victim's wet clothes off, get him or her into that sleeping bag along with another person that has also removed their outer garments. This body to body contact can help the victim get warmed and circulation restored. When a life is at stake, do not let false modesty delay your assistance. Warm fluids can start warming a victim from the inside but, if that victim is lapsing into unconsciousness, do not attempt to force down any warm beverages. If you choke them, you will not have to worry about hypothermia. In other words, common sense and care should be exercised.

As with many other problems, prevention of hypothermia is much better than the cure. Remember that to stay dry is to stay warm. When exercising you can regulate body temperature by using the layer system. A heavy outside garment is often too warm when exercising or too chilly when taken completely off, so wear several lighter layers of clothing which can be easier regulated according to the weather conditions and amounts of exercise. A chimney effect of moving air to carry away perspiration can be created by regulating the frontal and sleeve openings of the outer garments.

Keeping proper food in that fuel tank (stomach) is also important as a defense against hypothermia. Foods with a higher concentration of fat are recommended for winter use and will be explained in more detail later in the book. An important factor in maintaining the body core temperature is body shelter. This can be the clothes on your back or whatever is available in your pack, and anything that can be improvised from the immediate environment.

This now brings us to the subject of snow and how the rabbit utilizes it to conserve body heat. Snow is one of nature's most insulating materials if used properly. Let us take some lessons from the rabbit, ptarmigan, and other creatures that utilize the snow to maintain body heat. The ptarmigan, a small grouse like bird that is

14

a slate-grey color during the summer to blend with the rocks and alpine background, turns completely white during the winter for protective camouflage in the winter environment. The ptarmigan does not fight winter conditions, but uses them to it's own advantage. Utilizing the daylight hours, it feeds during the day on willow buds protruding above the snow in the high alpine terrain and rests at night. During a storm or at night, it fluffs out a small hole in the snow or takes advantage of an existing depression in the snow beside a willow bush and settles down into the hollow.

The ptarmigan utilizes the insulating properties of the snow.

There is no problem if a wind comes up and blows snow over the bird. With head tucked under it's wing, the ptarmigan can still continue to breath in the porous snow. The surrounding snow acts as insulation against loss of body heat. I have often traveled on snowshoes over the white snow surface during winter storms with no sign of ptarmigan or tracks. All of a sudden the snow explodes by my snowshoes and out flies a white bird. Snow had blown over the resting bird to form a protective blanket. When the crunch of my snowshoes came too close it burst from the snow with a small explosion. This indicates quite clearly how the insulating properties of snow kept the ptarmigan warm and how the porous texture of the snow still allowed the bird to breath. With this incident in mind, I began experimenting with snow as a shelter. The potential of such a shelter was limitless.

I was not satisfied with some of the types of snowcaves and snow shelter construction described in various winter survival manuals and magazine articles. After some trial and error attempts at the construction of the snow shelters and their insulating properties, I have devised an energy and time saving technic which I believe is far superior to the older standard small entrance method.

After experimenting during numerous winter trips, I have now become so confident with the use of snowcaves for body shelter that I carry no sleeping tent with me on cross-country tours into remote areas. I also discourage participants on my scheduled winter tours from carrying a tent. This adds weight to a pack and in turn takes away from the enjoyment of the excursion. The students sleep in snow shelters, with outside temperatures at 20 or 30 degrees below zero, and they find it much warmer than in a tent. When properly constructed, it will not freeze inside a snowcave with the outside temperature at 20 degrees below zero in a raging blizzard. You can put your water jugs, apples, and other perishables beside you on a snowbench, and they will not be frozen the next morning. I use the same light-weight sleeping bag on my mid-winter tours that I use during my high-country summer excursions and I have never slept cold during my winter outings.

Hopefully, the following chapters will help you to make more efficient use of time and energy when constructing body shelters during your winter excursions or emergency situations.

CHAPTER 2
SNOWCAVE CONSTRUCTION

Sufficient snow for snowcave construction can generally be found during the winter in most of the northern and Rocky Mountain States, although snow depth and season intensity can vary from year to year. The seasonal mild temperatures of fall and spring do not always justify the expanded energy and time of constructing a snowcave for body shelter. Instead, you can gather old stumps, logs, brush or any of mother nature's materials to form a suitable windbreak to sleep behind. Dry pine needles or grass can be placed under you for insulation.

If you have a suitable sleeping bag you can sleep comfortably with a bit of improvising during the early fall or late spring. With severe winter temperatures in a deep snow environment, you will be in trouble without some type of shelter. Therefore, learn to use the insulating properties of snow for body protection and well being. The first thing you need to learn is how to "read" the snow. The texture of snow can change from day to day and even from hour to hour. The type of snow makes a difference on how you undertake this project, but let's start with the type of snow most suitable for snowcave construction.

First, you should locate a deep, wind-blown snow drift that is not in a potential avalanche path, near an old tree that might blow down in a wind, or adjacent to any other hazard. The reason for seeking a wind-blown drift is that disturbed or blown snow is firmer than the deep fluffy snow under trees where the sun and wind has not compacted it. If possible locate your snowcave site on a gentle sloping hill. This will save you time and energy because, as you move the snow out of the cave cavity it will roll down the hill and out of the way. Otherwise you would have to continually stop excavating and shovel the snow away as it piles up in front of the cave.

If you have a small shovel with you, this will simplify the project. If not, you can improvise with a snowshoe, an auxiliary ski tip, a flat stick, can, cup, hub-cap from your vehicle, or other similar digging

17

tools. On a planned snowcaving excursion, you will naturally carry a shovel of some type which will make the task easier.

Tools for digging a snowcave can vary greatly.

Most old traditional methods of digging snowcaves are made with a small entrance hole which can create problems during the construction process. When making a small entrance hole to the snowcave you are continually brushing the top of the doorway and getting snow down your neck which can be most uncomfortable in cold weather. While digging in close quarters your knees gradually get wet from kneeling in the snow. This could develop into a serious problem especially if you have no way to build a fire and dry out. As you gradually make the entrance hole large enough to get inside and start excavating snow from the interior, the snow quickly piles up and blocks the small entrance unless removed.

To get away from the problem of wet knees and a damp neck attributed to the old method shown in some snowcave manuals I

18

prefer to open up a large hole approximately four feet high and about the same width, providing the snow is deep enough for this type of construction. Then I can stand in that hole with my digging tool and have room to swing or toss the snow completely out of the cave. As previously mentioned, with the small entrance type of construction, you push the snow toward the small entrance and then have to move it again as it piles up on the outside. Thus, you are conserving considerable time and energy by standing in the wide entrance and swinging that snow all the way out in one motion. The larger compact chunks can be placed to one side for later use.

A wide entrance makes for more efficient and faster snowcave construction.

Make a raised sleeping bench inside of wide entrance cave

Excavate out the interior to the desired size, depending upon whether it is a one person snowcave or for two or more. The recommended dimensions of a one-person snow cave is approximately four feet wide and sufficiently long to spread out a sleeping bag. For a two person shelter the side dimensions would be doubled. After excavating to the desired size, start the snow bench construction. This is a raised platform of snow a foot or more in height, and sufficiently wide and long enough for your sleeping bag. The principal involved here is that cold air flows down and the warmer air moves upward. When sleeping on a raised bench, it is warmer than sleeping on the floor of a snowcave.

To form this sleeping bench, shave the rough spots off the sides and ceiling of the snowcave. As smaller chunks of snow fall to the floor and are pushed over to the side to start forming the raised bench, that snow is being disturbed or moved. When this happens metamorphosis begins. As the disturbed snow is piled up and smoothed into the raised sleeping bench, it will soon harden and retain a permanent shape.

During this bench building process you must remember not to push down too hard on the loose snow or kneel on it because it would squash out from under you. After this piled up snow sets for twenty minutes or more you can then sit or lay on it, and it will be rigid enough to retain itself. With this hardening factor in mind, remember to smooth out the snow while it is still soft before metamorphosis sets in. If that bench is left lumpy or sloping, it is uncomfortable to sleep on. After getting into your sleeping bag and discovering a lump beneath your body, you can pound and smash it down with your elbow from inside the bag. It is simpler however to get it smooth in the first instance.

If you have a tendency to roll around in your sleep and possibly slide down into the trench, this can be easily prevented. After completion of the sleeping bench, simply scrape or pile a ridge of snow an inch or two high around the outside edge. This will harden in the same way as the disturbed snow in the bench. If you roll close to the edge in your sleep, that hardened ridge will keep you from going any further.

While building the raised sleeping platform, a trench of 12 to 18 inches wide should be dug from the entrance alongside the raised bench. The snow that is removed from this trench can be placed onto the bench to help make it higher. This furrow can be alongside a single sleeping platform, between double benches or across all of them on the lower end. If this trench is sloping downward toward the entrance, it allows the colder air to flow downward and out while the warmer air is retained in the cave dome. It also creates more height for you when moving inside and undressing.

The depth of this inside furrow and the height of the sleeping benches will depend on the depth of the snow where you are making the cave. If it is a deep hardened drift of snow, the cave can be made higher and wider for more freedom of movement. If it is fresh unsettled snow with no great depth, the bench cannot be as high and still allow space inside the cave for undressing or moving around. Because of snow conditions, I have slept on the floor of snowcaves rather than on a raised platform. By placing a backpack, boughs, plastic sheet, or some other object across the entrance hole, there is no movement of air in the snowcave to carry body heat away. There will be only several degrees drop in temperature if sleeping on the floors instead of on a raised bench.

As previously mentioned, the amount and texture of the snow will be the determining factor for the dimensions and style of the finished shelter. This and your own imagination or ingenuity are the limiting factors as shown by the sketches on page 35. It is very important that the ceiling or dome of the snowcave be left round with no flat areas. Like the Roman Arches, the rounded dome shape gains strength with added weight while a flat ceiling with no support will start to sag.

Do not leave rough or downward points of snow in the rounded ceiling! If there is any thawing of snow on the warmer dome, those droplets of water can trickle down the point of snow. Upon reaching that point, there is no place else to go, so the moisture will drip off onto your sleeping bag. With a smooth inside dome, any excess moisture would follow the outside down rather than dripping off in the middle. A word of caution, during the spring or other unseasonably warm conditions, do not leave your sleeping bag laying against the snowcave wall. It could become dampened if any moisture does move down the rounded edge. This melting snow inside the snowcaves is rare during the cold winter, but it can happen during warm spells or from too much heat created by bodies or cooking in the cave.

After completion of the sleeping benches and the trench, move in the insulation material, sleeping bags, and all other gear while the entrance is still large. As you spread these items out on the newly formed flat bench, be careful, as previously mentioned, not to exert any downward pressure on the soft bench. After smoothing out your sleeping bag or improvised insulation in an emergency situation, move outside the snowcave to start the next process of construction.

There should be some large chunks of snow scattered in the excavation pile which can be used along with those pieces already set aside for the sealing up of the entrance hole. Gather some of these firm chunks, square them up with a flat stick, shovel, machette, or other tool and start laying them across the wide opening of the snowcave to form a snow wall. As each block of snow is laid down, put a handful of loose snow on top of and in between each snow block as a bricklayer would put mortar between bricks. Metamorphosis will soon harden this loose snow and make the entire

Move in ground sheet and sleeping bags before closing wide entrance

wall rigid. When constructing this wall across the snowcave entrance put a mark at the top of the opening to indicate the center of the inside trench. This will later show you where to cut your final small entrance hole to match the inside entrance.

After some experience in shaping the snow blocks and placing them into a wall across the entrance of the snowcave, you will learn to completely close that large hole in a few minutes. Most beginners have a tendency to use small unsquare blocks of snow with not enough loose snow in between to harden the wall. If you have to use small blocks of snow, make the wall two blocks wide to make it sturdy.

When you start to construct this snowwall remember to build it far enough inside so that it will support the layer of snow forming the top of your snowcave. If this closing wall is too far outside in front of the cave and the ceiling begins to sag a bit, there is no support to stop the sagging. By constructing this wall under the cave opening, it comes up under the layer of snow forming the ceiling. This snowblock wall becomes very solid and sturdy after setting a few minutes. Then, if the ceiling above the wall does settle downward, it comes in contact with the top of the entrance wall and stops that downward movement. Thus, that snowblock entrance wall is an added support for the front end of the finished snowcave.

As you build this wall across the snowcave opening, do not bother to leave a small entrance opening, but build it as a solid wall. It is faster and easier than attempting to build blocks around a small opening and having them keep falling in. After that wall is completed, let it stand a few minutes as you gather fire wood or perform other camp chores. This allows time for metamorphosis to set up and harden the snow wall. After ample time has allowed for this hardening process, take your shovel, flat stick, or available tool and cut a small entrance hole at the bottom of that newly formed wall matching it to the scratch mark at the top so as to connect with the inside trench. Make it large enough for your shoulders and body to crawl through to the inside of the finished snowcave where your sleeping bag awaits you on the hardened bench.

Build a snow-block wall across the wide digging entrance.

Starting a snow-block wall across the wide digging entrance

Author stuffs last hole in wide entrance with snow block

25

After hardening, cut a small entrance hole in the snow wall.

Author exits from a hole cut in wide snowblock wall across snowcave entrance.

You can cut out a small shelf in the wall of the snowcave to hold a candle. After it becomes dark outside, that lighted candle will furnish all the illumnation needed and requires little oxygen. Place small sticks in the snowwall to hang up cups, caps, gloves,or other items that need to be handy. This is best done before you move in your sleeping bag so as not to scatter the fallen snow on your bed.

The temperature inside a snowcave is generally from 28 degrees to 38 degrees. It may be a bit colder when you first move in, but if you put your pack, a pine bough or other object across the entrance, your body heat will soon warm it to above freezing. This may sound a bit chilly to you, but there is no movement of air to carry your body heat away, so it will not feel that cold. If you are on a planned snowcave excursion, you would naturally have insulation pads, sleeping bag, and other items with you for comfort. During an emergency situation, you probably would not have these items with you and would have to do a bit more improvising.

In a cold environment or when sleeping on snow, approximately seventy percent of your body heat loss will be downward due to conduction with the ground or snow bench. So be sure and have insulation under you. During an unforeseen situation when you would not have standard insulation pads available, you can improvise with evergreen boughs, dry grass or brush. It is easier to put these items into the snowcave on the benches before you close up the large entrance. Then you don't have to drag them through the small entrance into the finished cave.

I generally don't advocate the use of evergreen boughs under your bed for ecological reasons, especially in heavily used areas. If everyone used them, the trees would soon be denuded and splotches of brown dried branches would appear all over when the snow melts. However they can be very useful in an emergency shelter.

If you snap off a few branches near the bottom of several trees instead of all removed from one tree, it will not be so harmful. In most areas, as these trees grow higher and thicker, the bottom branches will die anyway because the sunlight does not get to them. During an unplanned overnight stay in the woods, you most likely would not be in a heavily used area or near a trail.

Instead of just dumping these evergreen boughs in a flattened pile on the snow bench, start at the head and lay boughs on the bench with the courser tips slanting downward. As you work toward the foot laying on more boughs the fluffy upper ends of the boughs will cover the course tips and you will have a springy mattress. It can be two to six inches thick, depending on the availability of the boughs and how much time and energy you decide to extend into the task. The combination of pine needles and air under you makes for adequate insulation between you and the snow bench.

On some of my college student winter excursions where we attempt to teach conservation and ecology along with outdoor camping skills, I often give each student a flake of fluffy straw to take into the campsite. This hollow straw spread out under their sleeping bags makes a very good insulation. It weighs very little and is easy to carry if your camp is only a mile or two from the vehicle, and it scatters out as the snow melts in the spring with the grass coming up through it very well. Of course, you would not carry straw on long excursions, but this illustrates what can be done economically for scout or other youth excursions close to a roadway when special equipment is not available.

I DO NOT recommend air mattresses for winter camping because each time you move, the warm air under you is apt to be shifted and your body has to warm up more air. They are also generally heavier to carry than the foam pads. Rubber and plastic mattresses become very brittle in below zero weather and their use is restricted for winter camping. Air mattresses are also susceptible to puncture.

The 3/8 inch closed cell polyethylene pad is better for winter use than is a two inch polyurethane open cell pad and it is lighter to carry. The closed cell pad does not soak up water and gives more insulation between you and the ground or snow while the thicker, open cell pad will soak up moisture and allow more cold to seep through it. In all cases be certain that you have some barrier between you and the snow bench. Anything will suffice--pads, pine boughs, pine needles, grass etc.

If properly constructed, the temperature inside the snowcave remains the same whether the outside temperature is 30 degrees above zero or 30 degrees below zero. I have witnessed folks with normal outer wear sitting on the benches inside a snowcave and visiting for an hour or so in comfort. When removing their outer garments and preparing to slip into the sleeping bags, they seemed to be in no great rush because the outside cold does not penetrate into the snowcave.

The same relaxed atmosphere principal applies in the morning when they come out of the sleeping bag and start to get dressed. They are in no hurry and you do not feel the cold as when camping in a tent or under a lean-to.

Many of the snowcave manuals show a fire inside the snowcave. I CANNOT agree with this! I prefer not to cook inside the snowcaves for several reasons. To start with, if you have a fire or heat source in that snowcave you will need a vent in the top to let the gases out. With a vent hole in the top, the warmer air will also go out and will be replaced by colder air coming in the entrance hole. Without the vent hole, the warmer air rises and is trapped. If you are sleeping on the raised snowbench, you are taking advantage of the warm air.

A heat source in the snowcave will cause the snow along the walls to melt. If you happen to brush or touch a wall with melting snow, you will have a wet spot on your clothing. If there is no fire inside the snowcave, the walls will generally remain in the form of dry snow. If you touch them the snow will brush off your clothes but they remain dry. When the fire or heat source goes out, the dampness on the walls will freeze into a thin layer of ice preventing air from passing through the snow wall. You would then need a vent hole to allow fresh air to come in.

I DO NOT use a vent hole in the top of my snowcaves and have had no problems with lack of breathing air. Most snow is porous enough to allow sufficient air into the cave for several people, unless you have formed an ice layer from inside heat or are unfortunate enough to have a companion that insists on smoking in the snowcave. Then you would need a vent hole to allow that smoke to escape. Having a fire or heat source inside the snowcave raises the humidity and makes it feel colder to your body.

For many years while guiding college winter tours and other outdoor groups, I build or assist with the construction of 40 to 50 snowcaves each season. This has given me the opportunity for experimentation and I have come to the conclusion that no vent in the top of the snowcave is required. If you utilize a heat source in the cave or have smokers, a vent is required.

If several people have used a snowcave for several nights,an ice coating from the breathing vapors can form on the inside walls. This can easily be removed every three or four days if you so desire by first removing your sleeping bags or covering them with plastic sheeting. The thin ice coating can then be scraped off and the porous structure of the snowcave will again function normally.

If you are spending the night in a snowcave without the benefit of a sleeping bag and other comfort items, do not despair! It may not be the Waldorf-Astoria, but you will not freeze or get into serious trouble as you might if spending the night in sub-zero temperatures huddled under a tree or brush. Should you be without a sleeping bag in a snowcave you may not get a lot of sleep and you may get a bit chilly. However, you can use isometric exercises to warm up. Put your hands or feet together, push back and forth, rub briskly. Wrestle yourself so to speak and this will get your blood circulating again and you can then doze for a while. More on this later!

As previously mentioned, remember that snow is generally more fluffy and not as stable in shaded areas in the timber, while a wind blown drift extending out from a rock, brush, or some obstacle has a tendency to be much more compact. If you cannot readily locate a suitable wind-blown drift, there are ways of making that snow more solid before starting to dig. With a fluffy type snow or a heavy new-fallen snow, you can tramp back and forth over it while poking a stick up and down in it. This disturbs the snow and as previously mentioned, metamorphosis soon sets in and makes it solid. If the snow is shallow and you need more depth, the surrounding loose snow can be piled onto the area you have tramped. This rounded pile of snow will soon harden and you can then start digging your snowcave into it.

If all of the snow is only a couple feet or so deep, you can locate a sloping ridge that slants upward for ten to fifteen feet or more. Get on the upward side and scoot or slide the snow towards the bottom into a large pile which will soon harden as a result of being disturbed. You can use the same principal of piled snow on a lake or other level surface when needing snow depth for a shelter. Whenever using these piles of snow, it means more time and energy than when utilizing a natural drift of deep snow. You have to pile that snow up, wait for it to harden and then dig out the cave.

When snow is shallow, make a pile of snow for snowcave

After pile of snow has hardened a few minutes, start digging a snowcave

Completed snowcave in a pile of snow

The time of winter also has a great bearing on the snow texture. The snow is generally not as deep or as compact in the early part of the winter as it is later in the season when storm after storm has added weight and depth to the first layers of early snow. In the early winter when the snow is not of a solidified texture, you might have to constrict individual caves. A hole in the snow with a rounded dome about four feet wide for one person will not have a tendency to fall in as easily as would one that is six to eight feet wide. In the Rocky Mountains, the snow often gets ten feet or more in depth in the latter part of the winter and is very solid. We can then construct larger holes in the snow to accommodate six or more people with no problems.

It might bother some folks to sleep in a dark snowcave by themselves unless adjusted to it. For most campers, several people in the same snowcave adds appreciably to the overall enjoyment of the excursion. This allows for participation in evening discussions before or after the candle is blown out before going to sleep.

If the snow texture does not make it feasible for a large and wide snowcave, you can dig out an entrance trench with individual tunnels or holes about three feet wide for each sleeping bag leading off from this trench as per the cave design sketch. Leave a pillar of snow between each tunnel for added strength. With this design, each individual can get in or out of bed from the basic trench without disturbing anyone else. The participants can have that "togetherness" atmosphere in this design of snow shelter, but the snow cave is much sturdier because of the smaller narrower sleeping holes leading off from the center like wagon wheel spokes instead of one large and wide room which is apt to sag.

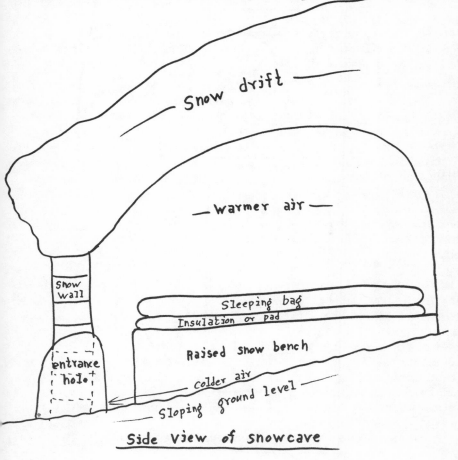

Snow drift ——

— Warmer air —

Snow wall

Sleeping bag

Insulation or pad

entrance hole

Raised snow bench

colder air

Sloping ground level —

Side view of snowcave

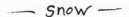

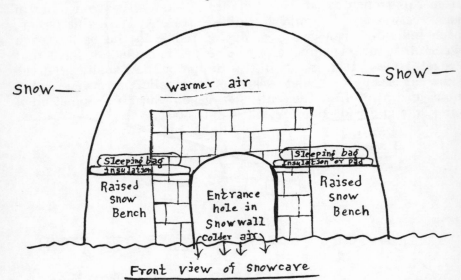

Front view of snowcave

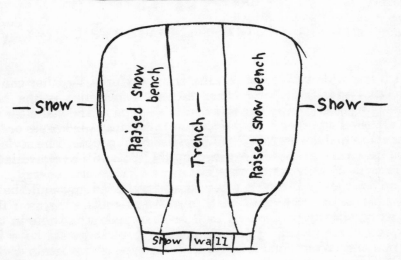

Top view of snowcave (2 man)

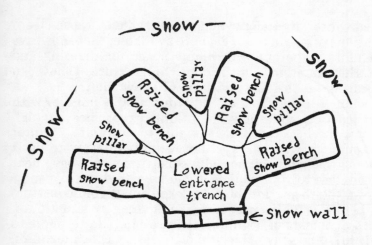

Individual sleeping holes for group when snow conditions do not work for one large wide snowcave.

With the wide variety of terrain, snow textures, weather conditions, tools available, and other factors, experience is the best teacher for determining the most suitable type of snow shelter for each occasion whether it be the narrow individual snowcaves or the larger more elaborate shelters to house several people. The style or size of snowcave is also determined by the tools you have available for the task as well as the circumstances of time and energy.

In an emergency situation where the output of energy could be an important factor combined with improvised tools such as a flat stick, cooking pan, or choosen tool, you can make that hole in the snow just large enough to accommodate you. It would be a bit cramped and inconvenient, but with some type of insulation under you, it would not take as much body heat to warm up that small hole as it would a larger cave.

On a planned winter excursion when a larger shovel, warm sleeping bag, insulation pads, and other gear are available, that snowcave can be much more elaborate with ample space for dressing and moving about. Keep in mind that the depth and texture of snow can also have a bearing on the size and type of snow shelter.

Aside from some tutoring from an individual that is familiar with snowcave construction and snow conditions, experience through trial and error is the best teacher. If unable to negotiate a winter snowcaving excursion with an experienced individual or group, I recommend some weekend you actually construct a snowcave near to your home or close to your vehicle if you have to drive to a snow area. This allows you to get the "bugs" worked out of your construction technics before actually needed in an emergency situation.

I have often seen a team of beginners require a half day or more to complete their first snowcave. They had to learn various technics, such as how to break loose the chunks of snow and roll them out by efficient methods. They learned how to regulate their layers of clothing and adjust their work pace so as not to create perspiration. After building several snowcaves I have seen these same teams construct a snowcave with sleeping benches and walled up entrance in an hour without even getting up a sweat. This learning experience could be very valuable at a later time during an emergency situation.

If you spend a night in that first snowcave close to your home or vehicle, you can check to see if your sleeping bag and insulation pad are suitable for this type of winter camping or if you should check into a heavier and warmer bag before attempting any extended winter tours. If things don't work out as expected this first night in a snowcave, you can retreat to the vehicle or home if closeby rather than being in a remote area with improper gear and experience.

I have found that beginners often have difficulty in determining how thick the remaining wall is when they are inside digging. To help solve this problem until more experienced, you can gather a few small sticks about twelve to fourteen inches long. Before starting to dig, poke these sticks straight into the snow along the top and sides of the snow drift or pile of snow. When you are inside digging and come onto the end of those sticks, you know you are close enough to the surface and not go any deeper. You will eventually learn to estimate the thickness of the remaining wall by the amount of light that filters through and then dig accordingly.

Check out your gear and technic ahead of time and be prepared for unforeseen situations. You can then have added winter camping enjoyment and comfort by being able to utilize nature's insulating material — snow.

The useful life-span of a snowcave made in a drift or bank of snow is usually three to five days before it starts to sag. This can vary according to snowcave size, snow texture, temperature, use of inside snow pillars, and other factors. Snowcaves made in piles of snow that have hardened from the piling process will usually last much longer. On occasion I have been back to snowcaves dug in piles of snow a month after the first use and have found them still in A-1 condition. After removing the snow blown across the entrance hole, I found no sagging had taken place inside the snowcave. I simply moved my sleeping bag inside onto the solid sleeping bench and had a good night's sleep with no expended energy normally required in construction of a new cave.

There is no problem when a heavy new snow falls on the pile of snow with a rounded snowcave in it, but it can create some potential hazards with a wide snowcave in a drift of snow. This several hundred pounds of extra weight could cause the roof of a wide cave to fall in while sleeping. Another reason why you should always dome shape the inside of your cave rather than having a flat ceiling as many beginners have a tendency to do. The narrower individual sleeping holes with pillars of snow between each opening as described earlier is much sturdier than a single wide snowcave.

This word of caution is not meant to spook you out of using snowcaves, but to alert you to conditions of added weight, natural cracks in the snow drifts, and construction design.

I have utilized dozens of snowcaves and have not had any of them fall in while sleeping in them. Experience will teach you to "read the snow" and use common sense accordingly.

38

CHAPTER 3
OTHER TYPES OF SNOW SHELTERS

Oftentimes, due to location or weather conditions, suitable drifts of snow for snowcave construction are not available. You then have to resort to other methods of shelter construction in order to utilize the insulating properties of snow.

Constructing an igloo with blocks of snow is another way of utilizing snow for body shelter. You have probably seen sketches in various publications illustrating nice round igloos made with large uniform blocks of snow. That type of igloo construction very seldom works in the snow texture of our Rocky Mountain Area and other states where the snow generally is not solid enough for large blocks. Those uniform blocks, approximately 18 inches wide by 30 inches long, can be cut from the hard-packed arctic snow which is of an entirely different texture than that of most Rocky Mountain areas.

You would need a snow saw or long knife to cut those large blocks, which is not likely to be with you during an emergency situation. It also takes a lot of experience and assistance to place the long flat slabs of snow on top of one another and have them remain there as they are gradually sloped inward to form the round dome as seen in the pictures of the Eskimo igloos.

I used to think an igloo was something the kids played with in the backyard after winter storms. But I came across situations where snowcaves were not practical, so I began experimenting with different methods of igloo construction. I am now using a snow block method that is more conducive to the texture of the Rocky Mountain snow cover. Instead of making the 30 inch snow blocks as shown in some of the manuals, I cut the snow blocks about 12 to 15 inches long and 10 to 12 inches wide and anywhere from three to ten inches thick depending on snow conditions and block location in the igloo.

Then instead of laying these blocks on edge, I lay them flat similar to laying a book on a table. There is then no problem of the blocks falling over as you stack them up into a wall. As you place these snowblocks into a wall, put a handful of loose snow between each block in the same manner in which a bricklayer uses mortar. As mentioned previously, whenever snow is disturbed, metamorphosis hardens the snow. Hence, this layer of loose snow between each block soon hardens to make a sturdy wall of snow.

Before starting to build the wall you will have to mark out a circle the size of the desired igloo and then place the snow blocks on this marker. Keep in mind that the larger the size of this base circle, the more difficult it will be to bring the top together and closed.

The square snow blocks are good for making a straight wall, but if trimmed to a slight modified wedge shape as shown in the sketch, they are more conducive to the circular wall construction.

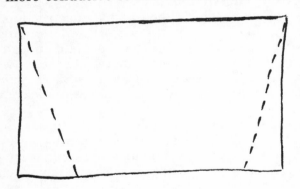

Igloo snow blocks are tapered to form circle.

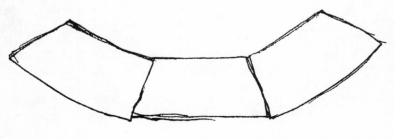

While building the lower layers of the igloo walls, DO NOT bother with making an entrance hole or doorway. It takes time to build around such an entrance hole and the blocks oftentimes keep falling down over that hole. So build a solid wall and a little later, after the disturbed snow has hardened, proceed to cut out an entrance hole as explained previously with the snowcave wall.

When starting to lay this first layer of blocks on the marked circle, you have several choices. You can cut blocks from within the inside of the circle or you can leave that snow and use it later for constructing the raised sleeping bench inside the igloo. If the snow is several feet deep, some of it will have to be removed before finishing the igloo.

Generally, a team of two people works very well for igloo construction, with one person cutting snow blocks and handing them up as the person inside the circle fits the blocks into the circular wall. When this wall gets several rows high, you should start offsetting or laying each block towards the center about two inches or more in order to start drawing that wall inward for the rounded dome. This is where the thickness of the snow blocks is important. They can be about ten inches or more thick on the lower layers of the igloo wall in order to make it come up faster. But as you build higher and begin to set each block in about two inches toward the center in order to close the top, you have gone up in height ten inches, while the diameter has closed by only several inches. If this continues, that wall would soon be higher than you can reach and still not drawn together at the top. By reducing the height of each layer of snowblocks near the top of the igloo and in setting each flat block toward the center, the rounded dome soon comes together. With the blocks being twelve inches or more wide, there is still plenty of flat surface for each block to lay on after being slid several inches toward the center.

When you get to the top of the igloo and only one large block is needed to close the rounded dome, it is then time to cut an entrance hole near the bottom of the snowwall. This will let the person inside exit before filling his or her neck full of snow when placing that last large block on top. After this last top block is in place, you can stuff handfuls of snow in any cracks that remain. Loose snow can also be applied on the outside surface of the igloo to add thickness or create a smoother finish. This snow will soon harden to make the entire igloo more sturdy.

Cutting and placing snow blocks for an igloo

Gradually place rings of snow blocks inward to form a rounded dome of the igloo

Cut entrance hole after igloo wall has hardened for a few minutes. Let inside man out before putting on top block

Stuff loose snow into any remaining cracks or holes in igloo wall.

As with the snowcave, you can construct raised sleeping benches inside if you desire them for slightly warmer sleeping rather than sleeping in the colder air on the floor. It is beneficial if the entrance is lower and sloping downward from the igloo floor so as to allow the colder air to flow downward and out.

If you are in an emergency type situation where the expenditure of energy and time is an important factor, make that igloo as small as feasible and still be able to serve the needed purpose. If you are on a regular winter camping excursion and have the proper construction tools, insulation pads, sleeping bags, and other gear, you can afford to spend more time and energy to make that igloo larger and more refined with sleeping benches, entrance tunnels, etc. Naturally, it would be larger to shelter several people than it would be for one individual.

I personally do not care to take the time and energy to construct the traditional snowblock-covered tunnel or entrance into the igloo as seen in most of the artist sketches. You can make that igloo entrance on the leeward side rather than facing the wind. This will keep the wind from blowing into the igloo. Once inside you can set your backpack or snowshoes against a piece of plastic or boughs across that entrance hole to detour the wind blasts if a shift of wind direction outside occurs.

More efficient tools are usually required for igloo construction than required for digging snowcaves. When a shovel is not available, a variety of items can be used to scoop out a snowcave as previously discussed, but to cut blocks of snow for igloo construction, a snow saw or long machete knife is ideal. I have experimented with a flat piece of aluminum sheeting and found it to be very efficient for igloo building. Go to your local sheet metal shop and get a piece of medium gauge aluminum approximately 14 by 14 inches, have one edge edge crimped one half inch and round and smooth the corners. The crimped edge gives you a hand hold and still allows the flat sheet to easily slip into a backpack and take very little space or add much weight.

To cut blocks, you can shove this flat sheet straight down into the snow and slide it across the snow to cut a long line. Cut another parallel line far enough apart to match the width of the blocks desired. Then turn the aluminum sheet at right angles and cut into the snow at ten to fifteen inches apart, or whatever length you desire the snow blocks. Now tramp down or remove the snow from one side of the row of cut blocks and shove the aluminum sheet horizontally into the snow from the side at whatever depth of block you desire (deeper for the bottom rows and less for the upper rows) and lift that block out. When the snow conditions are ideal, a lot of snow blocks can be cut with this aluminum sheet in just a few minutes.

This flat sheet is also very useful and efficient for the person inside the walled circle that is placing the blocks as they are handed to him or her. With the use of this aluminum sheet, this person can readily trim the corners of the bricks to fit the circular design, or slice the blocks in half for less height. Each winter camper can insert this aluminum sheet into their pack and never realize they have it as far as weight and bulk are concerned. As with any of the tools, some experience makes all of them more efficient.

Experience also teaches you what type of snow is suitable for snow blocks and what is not suitable. When the snow is not suitable for cutting snow blocks, you sometimes can change that snow texture. With your skis or snowshoes, tramp down an area and let it set a while for metamorphosis to harden the snow while you are gathering fire wood, obtaining water or other camp chores. After 20 to 30 minutes, which can vary according to temperature and snow conditions, you can start cutting blocks from the hardened area. Sometimes the snow texture will be composed of smooth round granules. This granulated snow will not adhere for snow blocks and will not work for igloo construction.

This is where the experience of "reading" the snow is very useful in knowing what you can and cannot do with it. Snow texture will dictate your decision of how to expend energy and time in the construction of snowcaves, igloos, or other types of snow shelters. When snow conditions are not suitable for snowcaves or igloos, limbs can be cleared out from a section under or beside a fallen evergreen tree. Brush and snow can then be piled around and over that hole to form a reasonably snug emergency shelter.

Limbs, snags, and brush can be piled over the edge of a large fallen log to form a cubby hole. Pile insulating snow over this framework and then put dry grass or evergreen boughs inside for insulation between you and the ground. Crawl inside from the open end and close that hole behind you with extra boughs and snow. This gives you some protection from the outside elements and helps to retain your body heat. Dry powdery snow that is not suitable for snowcaves or igloos can be utilized for this type of shelter.

Another type of snow shelter can be made by digging a trench about eight feet long and three feet wide in the snow. Pile the removed snow on each side. Metamorphosis will soon harden those piles of snow. Now you can lay skis, snowshoes, or snags of brush across the top. Next, cover it with finer brush and pile snow over the top as insulation. A poncho or piece of plastic is useful to spread over the brush on this type of shelter to eliminate snow particles filtering down, but allowing the first layer of snow to harden before piling on more snow also serves the same purpose. Move in boughs. grass or brush as protection from the cold earth. Crawl in the open end, closing it behind you with available material. You can wait for morning or the end of a storm without losing body calories to the colder outside temperatures.

A fallen tree provides the start of an emergency shelter

Lean snags and other material against the log before covering with smaller brush and snow

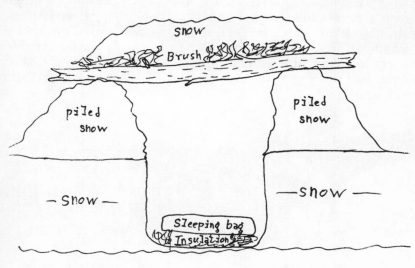

Snow

Brush

piled snow

piled snow

— Snow —

— snow —

Sleeping bag
Insulation

Trench Shelter when snow is shallow

The described improvised shelters on the preceding pages would be utilized for emergency and unforeseen situations when you do not have full camping equipment available. By improvising some type of body shelter, it helps to retain body heat rather than to lose great amounts of it by being exposed to wind and outside temperatures which can flitter away body calories quite rapidly.

During traveling or construction of shelters, do not forget that "to stay dry is to stay warm". As mentioned previously, some types of clothing, when wet, can lose up to 90 percent of their insulation value. When exposed to dampness and wind, you can lose body heat up to 24 times faster than in normal conditions. Therefore, pace yourself while making these shelters and keep perspiration to a minimum. Remember that your brain is still your best survival item during these emergency conditions, so think ahead and improvise accordingly. Rather than make a complete snow shelter during inadequate snow conditions, it might be more beneficial to get a warming fire started and use that in combination with the advantages of a lean-to improvised from available material.

The lean-to structures make efficient shelters when in timber or brush country. These can be made by propping or fastening a cross-pole about five feet high between two trees or other raised objects. Other dried poles or snags are leaned against that raised pole at an angle as shown in the sketch. Smaller brush and boughs are laid over this pole frame with an overlapping shingle effect starting at the bottom and working up. This makes the shelter more wind and water resistant.

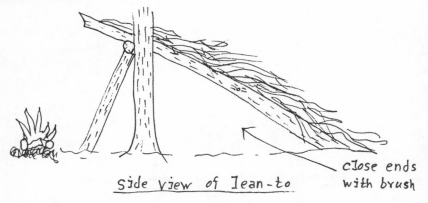

Side view of lean-to

close ends with brush

48

overlap boughs with shingle effect

Front view of lean-to

A fire is placed several feet in front of the lean-to. This small fire reflects heat to you under the shelter as well as radiating down on you from the lean to. If you place a pile of rocks or short logs on the opposite side of the fire, it will help to deflect more heat toward you. A wing of poles, brush, or boughs on each end of the lean-to helps to retain that reflected heat and deflect any side breezes. As with snow shelters, put some boughs or other insulation on the floor of the lean-to. Keep a small fire in front of the lean-to and you can have a comparatively comfortable night under the improvised shelter during cold conditions. It is recommended that you have a sizable pile of firewood gathered before dark that can be utilized during the long night hours.

If two or more persons are involved, one person can maintain a small, continuous fire while the other person sleeps for a couple hours on a dry cushion of boughs or grass under the lean-to. At the end of that period, they can trade positions with one keeping the fire going while the other sleeps. They are then rested when morning comes and ready to travel.

If utilizing a fire in deep snow country, you have to put something under that fire or it will gradually melt the snow under it and sink down too far to be of any use to you. You can usually locate some damp, rotten logs or green brush to lay under that fire before starting it. This will generally last for several hours before burning through because most of the heat goes up rather than down.

For some of the reasons previously mentioned, I DO NOT use a fire inside a snowcave or igloo as shown in some survival manuals. If you have digging tools with you, you can scoop out an area in front of the snowcave or igloo that is a suitable size for the number of people in your group. The removed snow can be piled into a semi-circle wall around the fire, which is built on the ground so that it does not sink into the snow as it burns. You can form a snow bench about eighteen inches high around the bottom of this semi-circle of piled up snow, and cover it with dry grass, boughs or a poncho. After a few minutes when metamorphosis hardens this disturbed snow, you have a sturdy bench to sit on by the fire. The wall of snow around you deflects any wind, and much of the heat generated by the fire is retained or reflected within that circle.

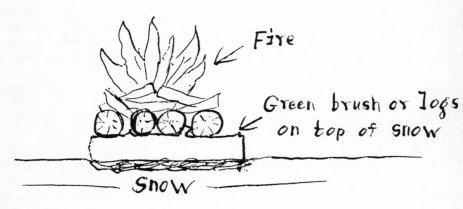

Insulation between fire and deep snow

Top view

Side view

Gathering area for groups outside the igloos or snowcaves before retiring for the night.

After your cooking or campfire discussions, you can then retire to your snowcave or igloo for sleeping. Inside, the temperature may be chilly, usually around freezing, but there is no movement of air. Because of this "still air" condition that does not blow body heat away, you will not have an impulse to rush while undressing before getting into your sleeping bag. It does not take long to be warm and cozy inside your sleeping bag with the slightly above freezing temperature inside the snowcave or igloo. It would take a lot longer to get warm if you were outside, in the zero or below temperatures with a wind blowing. That outside wind can filter away a lot of body heat that is better retained in the still air inside a snow shelter.

If you do have a tendency to remain cold inside your sleeping bag, it may be due to poor body circulation. While waiting by the outside campfire or eating, you might have gotten chilly because of lack of body activity and begun shivering before going into the snow shelter and crawling into the sleeping bag. A sleeping bag does not create heat, but retains it. If the circulation has slowed in your body extremities such as hands and feet leaving them with a cold feeling, correct the problem before retiring to the sleeping bag.

Before going into the snow shelter, jump up and down, push your hands and feet against each other, run in place, or any other type of action to get that circulation going again. After your body has warmed from the exercise, undress and crawl into the sleeping bag. You will then sleep warmer. If not desiring to waste energy, take off your shoes or boots and get into your sleeping bag. Then proceed to take off trousers, sweaters, or other outer garments while inside the bag. During all of this squirming and wrestling around in the confined quarters of the bag, your body and the sleeping bag have warmed up and you will drop into a comfortable sleep rather than laying there and shivering because of poor circulation. If your feet have a tendency to get cold, put on a pair of fresh dry socks.

When in an area where there is abundant firewood laying around and where there is not a lot of human traffic, I like to utilize a small campfire in a prepared area as described in the sketch on page 51. The crackling sound and the reflected heat of the evening campfire in the protected area just outside the sleeping shelters have a tranquil effect as well as helping to keep your body warm before retiring. Before leaving the winter campsite, I scatter the remaining small amount of black ashes so as not to leave an eyesore when the snow melts.

Many backpackers and outdoor users now carry the small backpack stoves with them. There are many types and brands available depending an individual's intended use or desire, but keep in mind that not all campstoves are suitable for winter use. Some of the gravity type stoves and those utilizing a vapor from heated fuel generally do not function properly in extreme cold. For winter use, I have had good results with the Coleman Peak I stove which uses white gas. There are other brands available that function efficiently in cold weather. Check them out and then make your choice. A trial run or checking with a person that has had experience with the stoves in cold weather is recommended.

If you are utilizing a heavy use area for your winter camping where firewood is not available or ecologically not advised, small stoves are very useful. They can be used in moderation inside the snowcaves or igloos without creating any extensive humidity or melting snow problems as might develop when using a fire inside the snow shelters during storms or extremely cold outside conditions. These small backpack stoves can be utilized for heating water for hot beverages and cooking simple meals inside the snowcave.

During winter cross-country tours on skis or snowshoes, these stoves, bottled fuel, tents, and other accessories all add more weight to the pack which takes enjoyment away from the excursion. After many years of learning to improvise with what nature has available, I prefer to go light and enjoy myself. I also find a personal pride in making myself comfortable with the available material. If you are going on a planned winter excursion, you can make yourself reasonably comfortable by carrying certain items with you and then utilizing the natural materials available in the area. Experience and a knowledge of materials available in your use area for improvising will determine what you elect to carry and what items you will eliminate from your pack. If you are caught in the woods during the winter in an unplanned situation without the desired equipment, learn to improvise body shelters from hollow logs, holes in the snow, and other means described earlier. Look around and use your brain to remain comparatively comfortable with what is available.

CHAPTER 4
WINTER CLOTHING

It is most important to start out with proper winter clothing, in addition to learning how to improvise body shelters from snow and other materials that are found in nature's domain. If a delay happens due to weather conditions, a mechanical breakdown, disorientation, or whatever, you will be better prepared to cope with the situation rather than having to rely totally on what nature has available.

Cotton clothing is NOT recommended for winter use. When wet, it has an osmotic effect that can get you into real trouble. If you were to get your trousers wet or damp by stepping through the snow into a hidden fresh water spring, by kneeling or walking in the snow, or by other means, that wetness would continue to spread throughout a cotton garment. The normal evaporation from damp cotton cloth will take a lot of body heat with it.

When wearing wool garments, your trousers would be wet only to the water line and that moisture would not spread. Wool also retains some insulation properties even when wet, whereas damp cotton loses most of its insulation value. If your clothes suddenly become saturated with water from breaking through thin ice or slipping, immediately roll the wet parts in fluffy snow. This snow quickly absorbs some of the moisture from the wet garments similar to cornmeal or flour soaking up grease or water. Before it freezes, quickly brush off the snow that has absorbed some of the moisture and brush on more snow to absorb still more of the moisture from the woolen garment. By repeating this process several times you have removed a large percent of that moisture before it freezes.

After wicking away the excess moisture from wet garments with snow, remove wet socks before they freeze, and put on dry ones. These dry socks will absorb moisture from your wet feet as well as from your boots. This will allow you to continue to travel to camp if it is not too far. However, if camp is too far for the prevailing weather conditions, build a fire and dry out before moving on.

During extended excursions into remote areas, I do not bother to carry extra trousers or shirts with me because I do not desire to be burdened down with the excess weight and bulk. ALWAYS carry extra socks, gloves or mittens, (preferably wool) and a fire-starting kit. I do not mean just matches, but a reliable fire-starting kit that will be described later.

I recall one incident when I had snowshoed into a remote area during the winter to run a trapline for bobcats and marten. During the day I became completely engrossed with setting out traps and snares for the fur-bearers, and darkness caught me on the side of a steep hill with a snowstorm moving in. In the darkness it took me an hour or so to follow the game trail down off the mountain. Upon reaching the bottom of the canyon, I soon located a reasonably flat spot to set up camp behind some spruce trees out of the wind.

Since I had not crossed any open springs for drinking water during the day, I was thirsty. The intake of fluids during cold weather is VERY important. Therefore, before starting to set up camp in the dark, I grabbed the "billy can" and a cord from my pack, left my snowshoes with the gear, and wallowed through about two foot of snow towards the Los Pinos River which was only fifty yards away. While easing along a flat spot looking for a suitable location to toss the "billy can" with the attached cord out to the flowing water to fill it — Bingo, It Happened! The snow suddenly gave away under me. I started sinking into the icy clutches of the Los Pinos River. As the icy water enveloped my legs, numerous thoughts began to race through my mind. How deep is the water! How will I get out! Will I freeze to death from being wet! Will anyone find me! It is amazing how many past years and circumstances flashed through my mind during that instant.

There was not a human within ten miles of me, so amid the many fleeting thoughts, I realized survival was up to me because no one else would help. I immediately threw myself backwards, and kept stroking and fighting backwards as the snow-covered ice continued to give way under me. During my struggle backwards, I soon reached a willow bush growing along the bank. I grabbed onto the woody limbs and pulled myself out of the icy water and onto the bank.

Being completely wet from the waist down, I immediately rolled in the snow to help absorb some of the water from my clothes. After

56

brushing off the water soaked chucks of snow, I wallowed through the snow back to where my gear lay, breaking off several small dead aspen trees protruding through the snow for firewood. After tramping out a spot in the snow and grabbing the fire-starting kit from the side pocket of the pack, a fire was soon going. With snow coming down in big flakes and small dry tinder not readily available in my haste, this would have been no easy task with standard matches. But with my reliable fire-starting kit, a crackling flame soon began to throw out both heat and light. This again illustrated to me why a person should always carry a good fire-starting kit, and why you should keep it where you can get to it immediately.

There was not enough firewood gathered to burn very long, so before settling down by the welcome warmth of the crackling fire, I sloshed through the falling snow to gather several more armloads of dry wood. By now my pant legs were frozen stiff and the water in my boots sloshed around with each step, reminding me of my serious situation as I broke off small, dead, aspen trees sticking out of the snow.

My reason for gathering more dry firewood instead of staying by the fire was simple. Standing by the fire, the heat would begin thawing out and drying my wet clothing. This warmed clothing would immediately become wet again as the snow melted upon coming in contact with the warm surface when venturing out for more wood.

With my trouser legs already frozen, it seemed to form a wind barrier and kept the loose snow from melting and penetrating, so it took only a few minutes to gather a substantial pile of firewood. I also gathered a pole and some branches to form a lean-to with the aid of a space blanket from my pack. Then I sat under it next to the fire and took my wet boots off. When going into remote areas by myself, I always carry an extra pair of light mukluk boots in the pack along with several pair of wool socks. Upon taking off the wet boots and socks, and drying my feet, I put on dry socks and dry mukluks. After a while in front of the fire, my pant legs became almost dry. Then I proceeded to dig a trench in the snow under an evergreen tree, and layer it with some pine boughs. After laying out my insulation pad, sliding the sleeping bag into the Gor-Tex bivouac cover, the bed was ready to crawl into. For various reasons I did not take the time and energy to build a snowcave that night. Some branches were laid across the snow trench under the tree and covered with smaller boughs and brush. This made a warm hole to sleep in and it was out of the wind and storm.

After completing the essential chores I cooked a hot supper and ended the meal sitting under the lean-to by the glowing embers of the campfire and sipping hot cocoa. Finally, the last moisture dried out of my pants and socks so that they would be ready for in the morning. The embers of the fire, with an occasional refueling, talked back to me. In the background was the soft murmur of the falling snowflakes as they sifted down through the trees. I have always enjoyed the conclusion of the day by the campfire during my outdoor excursions. This evening seemed extra special after having mastered a very serious problem which could have ended in tragedy.

With no human within ten or fifteen miles, it gave me an inner pride to know that I had coped with this situation without outside assistance. I felt relaxed and confident amid the refreshing mountain air and would sleep soundly in my snug and dry sleeping bag. As for drinking and cooking water that evening, I melted snow. I had already made one mistake earlier, and was not about to make another trip to the river to fill the "billy can" with water.

If you are traveling into remote areas for pleasure excursions, or work errands, it is best that there are two or more persons in case of some unforeseen situation. However, due to my past experience as a trapper spending much of my life in the mountains by myself, I am often alone in remote areas. I enjoy it, even if it is not recommended although my experience helps to add security and confidence. I can travel over the snow in the moonlight at night, or sit and watch a beaver or deer for an hour or so during the day and no one is there to think I am crazy for wasting my time. I personally think the human critter is backward in the out-of-doors. Deer, elk, coyote, beaver, mice, and other critters do most of their feeding and roaming at night while man has gradually developed the habit of stopping and making camp soon after the sun goes down.

Think about it a bit! When you are moving, body heat is being created from the friction of body muscles. If you do not have a dry and warm sleeping bag, dry clothes, or other suitable body shelter and attempt to lay down and sleep during the cold night, it will be a long and miserable night unless you know how to improvise from nature's available materials. If you know your terrain, have a moon or starlight to travel by, and other factors, you could continue to travel at night if it is an advantage to do so. You would continue to stay warm from the movement of body muscles. During the day, you might come across a nice, sun-warmed rock out of the breeze where you can stretch out and catch several hours of sound sleep. This is a lot more comfortable and much more restful when compared to shivering under a tree at night without proper equipment. Of course, this is strictly individual opinion and you have to weigh the pros and cons regarding your situation and the present circumstances. There are some actions that only experience can imprint into your brain for future reference and benefits.

For example, you might fall through the ice on a lake, creek crossing, or hidden spring. After you have gotten out and are rolling in the snow to absorb the excess moisture, your partner, if available, can immediately go to a suitable location with wind shelter and firewood available and get a good warming fire started. You can then dry your clothing, either by turning back and forth by the fire, or by hanging your outer garments on sticks or branches by the fire. As your clothes are drying over the fire you can stay under an improvised lean-to built next to the fire. This way you can absorb heat and finish drying out your underclothes. At this point the normal tendency would be to stop and make camp at that spot since you are partially wet and uncomfortable. This is a mistake, both physically and mentally. When you are dry enough to travel and, if there are several hours of day-light left, pack up and start traveling along your intended route.

Nothing warms the body as much as physical action. As you continue to travel, your mental attitude also changes because, instead of sitting around a fire and thinking about the mistake you made and time lost, you are moving on and realizing that it did not bother you that much. When the end of the day comes along and it is time to set up camp, you can go about this customary chore with a good physical and mental attitude.

This brings to mind an incident several winters back when my partner and I were hiking at night up a mountain valley with a creek winding down through the grass meadows and willows. A full moon was shining on four inches of new-fallen snow, making visibility sufficient for night travel. Ice lined the edges of the creek as open water ran in the center of the stream.

We were trapping beaver and muskrats, so we both had on hip boots in order to wade the numerous stream crossings enroute to our destination. As my partner, John, eased into one of the stream crossings which was about knee deep, he slipped and fell down into the icy water. He sucked in his breath as the cold water penetrated his clothes. In his haste to get up and, with the heavy pack on his back, he floundered around and lost his footing again on the slippery, moss covered rocks beneath the surface. This time he fell all the way down into the water and got completely soaked.

As he clambered out onto the creek bank I immediately pushed ahead through the grassy meadow for fifty yards to a grove of spruce trees which would give us protection from the wind. Brush and dry wood were gathered to get a fire started. Heat was starting to radiate from the crackling fire as John shuffled over with his clothes already stiff from the cold. John took off his frozen outer clothes and we propped them up on willow frames by the blazing fire to dry. While they were drying, I kept gathering dry wood to keep the fire going. John kept moving back and forth by the fire and turning first this side and then that side in an attempt to get his long johns dried out.

Considering everything that had happened it made a humorous sight. To make it even more impressing, a coyote on a ridge above us started to howl in the moonlight. I told John, "That coyote is laughing at you and your funny method of getting dry". He grinned back with a knowing nod.

After getting John's clothes dry from the reflecting heat of the fire, we continued on our way in the moonlight. Besides keeping our bodies warm from the movement, the activity kept our mental attitude positive.

This is another example of why you should always have a dependable fire starting kit readily available. When looking back to these occasional situations that could have been serious, you can get a chuckle from them along with stored knowledge for future use.

In addition to using the proper type of winter clothing, as mentioned earlier, learn to utilize the layer system which can be regulated according to exertion and weather conditions. If you have only a heavy down jacket, it can become too warm to work in, if you take it off, it is too chilly. However, if you have several lighter sweaters or jackets, you can solve the problem by adding or removing clothes as needed. Wool has excellent insulating properties, even when wet, but it is too itchy for some people to wear. For these folks, the Duofold type underwear is good because the inner cotton lining can wick away the perspiration, while the wool and nylon outer layer retains the heat as it breathes.

There is also "fishnet" underwear where quarter inch holes appear to wick away all of the body heat. However, these holes actually create air space in the openings of the net. A sweater or wool shirt worn over the "fishnet" underwear helps to retain the dead air spaces and still allow moisture to escape. Hence the "fishnet" underwear is quite warm in spite of its looks.

Polypropylene underwear is excellent for outdoor activities. Perspiration readily passes through to your outer layers of clothing keeping you dry and comfortable. I have been using the Damart underwear for some time and have had excellent results. This material, Damart Thermolactyl, is soft and smooth next to the skin, does not retain body moisture, and has very good wickability. This provides an excellent base for my other layer-system garments. You can write to Damart Thermolactyl, 1811 Woodbury Avenue, Portsmouth, New Hampshire 03835, for a catalog containing a wide variety of winter garments. The R.E.I. catalogs and other supply firms listed elsewhere in this book are also excellent sources for winter clothes including polypropylene garments.

The combination of wool shirts and loose-knit sweaters, and a jacket can be topped off with a loose-fitting wind parka of tightly woven material. This parka should have a hood and, by being loosefitting, it acts like a shell during windy or stormy periods. Winter campers actually need two sets of clothing which can over lap. A lighter set is needed for active use on the trail when you are producing body heat through movement of muscles. A heavier set of clothing is needed when the camp chores are completed and you are relaxing by the campfire or in the shelter. This wearing apparel can be used interchangeable and will vary with each individual according to your own body metabolism. Everyone's heat production varies according to many different conditions. Therefore you have to work out a clothing system best suited for you in accordance with terrain, type of weather to be encountered, length of trip, and other factors.

I have found that many folks new to winter camping have a tendency to wear too many clothes, especially when traveling. In addition to having heavy jackets for the purpose of insuring that their upper bodies stay warm, they often have heavy down trousers or similar leg gear. When snowshoeing or cross-country skiing, especially with a pack, they soon become too warm and begin to perspire, but cannot easily adjust the clothing as needed. Their underclothes become damp, which is no great problem for a one-day tour that terminates in a heated lodge or in a heated car with a quick ride home. But if you are spending the night out in sub-zero temperatures, damp clothing can develop into a serious problem.

When I become too warm during winter travel, I tend to lose a lot of "zip". Even in below zero temperatures, I often travel in just wool pants and shirt over my underwear with my head, hands, and feet protected by proper gear. When traveling at a normal pace, my heat production keeps me comfortably warm. If I stop to rest or eat, I can then put on another layer of clothes from my pack during any period of inactivity. I can control body heat and perspiration accordingly and have energy left at the end of the day. I have a much better tolerance for cold conditions than a lot of my companions do. I do not seem to lose much body heat through my legs, so I never bother to carry extra trousers with me. But I do carry some sort of a windbreaker, for my legs. Rain chaps are a basic part of my equipment.

Other factors, such as body conditioning, are involved. I am adjusted to physical exercise as a result of putting in day after day of traveling over the snow while conducting winter tour groups or while running a trapline on snowshoes in the mountains. My companions have spent much of their time setting behind a desk and do not completely adjust during a once-a-year excursion. All of these factors have to be considered when choosing your winter clothing as dictated by your own metabolism and other physical needs.

Consider your mental attitude. If you feel you are going to have a cold and miserable trip, you probably will. Develop PMA (Positive Mental Attitude) for better results regardless of whether you are using it for recreation or business contacts. Positive mental attitude is important for your better well-being. With many of us living in a partially controlled atmosphere, it is often difficult to establish a positive mental attitude during adverse conditions on outdoor excursions. This is where experience with combination of proper clothing and gear is of a great benefit in maintaining the proper attitude. Since comfort is an important factor towards your mental attitude, you now realize why clothing is a major part of your outdoor tour planning.

I have discussed various types of underclothing and would like to make some suggestions on outer clothing for winter excursions. You can lose more body heat through your head than through any other part of your body, so plan to wear proper headgear. Some folks like the knit caps that can pull down over the ears and neck when desired. These work very well inside the hood of a parka. Others like a cap with a bill on it for added protection against the sun or falling snow. A face mask or balaclava can be carried to use in extreme cold conditions in order to keep frostbite off of your nose and cheeks. Personally, I prefer a long knit scarf that can be wrapped around the face with the ends tucked under your outer garment. This gives added protection for the neck and upper chest. The scarf can easily be removed when the wind or blizzard conditions pass by. A word of caution! If using a snowmobile or motorized vehicle, be careful about letting loose ends of a scarf dangle. These ends can catch in the gears or belts and suddenly entangle you - need I say more about what might happen?

I often see people remove their caps or head gear during the day when skiing or snowshoeing. This can rapidly release a lot of body heat which means you are allowing body calories to flitter away. Attempt to guard against this careless expenditure of body heat and energy so as to have some reserve to utilize if an unforeseen situation develops. Keep that cap on and regulate your body temperature by utilizing the layer-system of clothing and adjusting your pace accordingly.

If you have loose fitting garments rather than tight, you can easily regulate body temperatures according to exercise and air temperature changes. This can be done by loosening the clothes at the neck, cuffs, and waist. By letting air into the lower openings at the waist and cuffs, the heated air has a tendency to travel upward toward the loosened neck. As this air escapes upward in a sort of "chimney" effect, it carries excess heat and moisture with it. You must think and act ahead to maintain the efficient body temperature. Start loosening outer garments when the body heat begins to build up and not when perspiration becomes a problem. If you wait too long to loosen your clothes, the evaporating perspiration will cause you to cool off too fast and you will become chilled.

Keeping feet warm is a major problem with most winter campers. Every experienced individual has his or her own pet way of solving this problem. For example, I do not like insulated boots found in the average shoe store because they make my feet cold. They are usually heavy and bulky and when I walk in them, my feet go plop-plop or tramp-tramp with very little action on my feet inside the stiff boot.

I prefer a mukluk type boot made of several layers of light and pliable material. As I walk my toes and feet are moving and stretching with each step which keeps the circulation going. I can also cover more miles per day with the light footgear than I can with a heavy boot because I do not get tired so easily from lifting a heavy boot with each step.

The rubber-bottom pacs are available in many of the sporting goods stores and are popular with hunters and winter users. This type of boot has a tendency to retain moisture from perspiration and they will result in cold feet. Whatever type of boot you choose for winter use, they should be loose enough to allow a liner to fit inside. Besides adding insulation, this liner also absorbs moisture from the foot and can be pulled out of the boot each evening and allowed to dry.

Felt liners are inexpensive and available at most sportsman's shoe stores, but they should be taken out of the boot and dried every evening. The Polarguard booties are good because they let some of the moisture escape. I make my own mukluk liners by hand-sewing natural sheared sheepskin into a sock shape with the fleece side in. I like to have the thickness of the fleece about one half inch. They are warm and wear well. This type of tanned hide is generally difficult to locate and it takes some sewing experience to get the correct shape and size.

An insole or two as added insulation between your foot and the cold ground is also recommended. Felt insoles usually absorb moisture while an insole made from a Polarguard batting will allow moisture to escape to the outside.

Whatever type of boots and liners you choose, make sure that they are loose fitting in order to allow for the extra insoles and socks. A tight boot restricts circulation which helps make your foot cold. If the boots are a bit too loose, this can be corrected by putting on another pair of socks, but if they are too tight, there is nothing you can do to correct the problem. Always carry extra socks with you during winter trips in case your regular socks become damp from perspiration or other moisture.

Speaking of boots, I would like to suggest that you check the American Waterproof + Plus boots made by American Footwear Corporation, One Oak Hill Road, Fitchburg, Mass. 01420. A Gore-Tex inner boot keeps your feet dry from outside moisture while letting perspiration filter away. I have found them to be extremely efficient for my outdoor excursions.

If hiking or working in deep, fluffy snow, you will probably end up with damp or wet shoes, socks, and pants legs unless you wear gaiters. Gaiters are a short wrap around type legging that fastens over the boot top and bottom of pants with snaps, velcro fasteners, or a zipper. This keeps the snow from sifting into your boot top and then melting from the body heat. A strap or cord from one side of the gaiter under the instep and tied to an eyelet in the gaiter on the other side will keep it from working upward and allowing snow to filter into the boot top.

Since the gaiters only go up to the knee or just part way up, some folks prefer to wear rain-chaps over the gaiters. This is a legging of light-weight, but water resistant material and slips over each pant leg and fastens to your belt. Make sure these are loose fitting to allow some movement of air to carry perspiration away. To me, gaiters and rain-chaps are a must for my winter excursions.

Gloves, like footgear, are a matter of personal choice and needs. I prefer to wear gloves with mittens over them. The mittens can be removed when more dexterity of the hands is needed, then replace the mittens to retain heat. Carry several extra pair of dry gloves or mittens with you on winter excursions. You can switch back and forth as necessary when one pair becomes wet or damp.

There are several things you can do as you make camp at the end of each day to help keep your clothing dry and warm. When you arrive at your campsite and get your snowcave constructed, tent set up, or whatever, get in the habit of doing all camp chores, such as wood gathering and getting water, before settling down by the warm fire. There are several reasons for this.

First of all, your body is warm from the activity of preparing camp. While still warm and active, but before slowing down and putting on that heavier garment from your pack, wade through the snow to gather enough firewood to last all evening instead of just enough for a few minutes. If your water source is nearby, utilize it instead of melting snow, which takes extra fuel. Gather some limbs or twigs to be utilized around the campfire as drying racks for socks, mittens, and other camp garments. When all of these camp chores are completed, then you can relax by the fire, take off your damp socks and put dry ones on. If your trousers or other garments are slightly wet, start drying them by the campfire heat while cooking the evening meal. When you once get dry and comfortable, you can stay that way until crawling into the sleeping bag for the night. As previously explained if you had only partially completed the wood gathering and other jobs away from the fire and then had to leave that warmth to go out into the snow again, you have a chance of getting damp all over again. Your clothes and boots are warm from the campfire heat and when they come in contact with the snow, that snow will melt and soak in very quickly. You might get cooled off while searching in the dark for more wood or frame limbs, and then you would have to get warmed up and comfortable all over again.

By completing all the chores before settling down for an evening by the fire, you can then fully enjoy this choice part of the day while cooking, drying out boot liners, damp gloves, or other items. The glowing embers of the campfire sort of talk back to you in a warming and soothing manner, and it makes for a comfortable and relaxing evening.

You will probably find it advantageous to give some attention to your boots before you retire for the night. After being exposed to the snow all day, your boots may not be completely dry when you are ready to go to bed. Instead of leaving them in a crumpled heap beside your sleeping bag, make sure that the laces are loosened, the boot is opened up, and left standing up straight. In the morning when that damp leather might be frozen and stiff as a board if sleeping in a tent, you can still slip your foot into that boot. As your foot warms up the frozen leather through the socks, that leather will soon thaw and become pliable so you can walk comfortably. This would be much better than not even being able to slip your foot into a folded or crumpled boot that is frozen stiff. This is more important if you are spending the night under a tree or in a tent instead of in a snowcave, because the temperature outside would be a lot colder than the temperature in a snowcave or igloo where your boots would not freeze stiff.

Slightly damp socks can be laid beside you in the sleeping bag at night and will generally be dry in the morning. If your feet have a tendency to become chilly at night, keep a pair of clean woolen socks available to slip on each night before retiring to the sleeping bag. It makes a big difference in sleeping comfort. If several fist-sized round rocks are available, you can heat them in the campfire, slip them into a sock or similar and put them in the bottom of your sleeping bag to warm your feet. A word of caution, don't allow them to get too hot to damage your sleeping bag or burn your feet. Do not use rocks from a stream or wet soil because the water in the damp socks could create steam and explode.

The sleeping bag must be given primary consideration for winter camping. Are your winter outings going to be weekend tours of two or three days or will they extend into a week or more? This is very important and can have an impact on your overall comfort and enjoyment. Down filled sleeping bags are usually preferred and are generally used by winter campers. Keep in mind that when downs get wet, you have no insulation value except the nylon covering on the bag. It also takes a long time for down to dry out after it once gets wet. During summer excursions, the down sleeping bags can be exposed to the air and sun for a few minutes each day to wick away moisture, but during the winter, problems can develop with a down-filled bag.

A human body loses about a pint of moisture in the form of respiration and evaporation during a night's sleep. Some of this moisture has to filter through the sleeping bag. If you have ever covered your sleeping bag with a sheet of plastic when retiring, you probably remember the next morning how wet your bag was under that plastic sheet. The moisture had filtered through the sleeping bag, but could go no farther because of the plastic barrier. When you are sleeping in a cold environment, this moisture filtering through the sleeping bag can create problems. Vapor filters through the warm part of the bag next to you but, as it drifts through to the colder temperatures toward the outside of the bag, it reaches the dew point and condenses to form tiny ice crystals. Condensed moisture is retained in the tiny down filaments which, after each night's use gradually become more soaked and frozen. The down bag begins to lose its heat retaining properties and the down filaments do not readily release moisture even when exposed to the sun or a wood fire. When this condensed moisture builds up night after night, the bag becomes damp, frozen, and virtually useless after a week or so of continual use without drying.

A down sleeping bag is fine for summer use or for short winter trips, but it can create problems during long extended winter tours. Check into sleeping bags containing fill materials of Polarguard, Dacron II, Hollofil II, Quallofil, and other man-made fibers. These materials are a bit bulkier and heavier, but have other advantages. Dacron II and Hollofil II are Dupont products with a short and slippery fiber that requires stitching between the inner and outer linings to hold the fibers in place. Polarguard is a continuous filament Celanese fiber that comes in a mat and would not require as much stitching to retain its shape. The Polarguard, the Dupont fiber fills, or other synthetic fills will not absorb moisture like natural down. If a fiber filled bag or garment gets wet, it is the lining that is wet and not the fill fibers as they absorb very little moisture. It can readily be dried by exposure to the sun, a breeze, or even a campfire (with attended caution) while down is very difficult to dry during an extended winter outing.

Perhaps you already have a down sleeping bag for your summer trips, but it is not quite heavy enough for winter use. You can solve this moisture problem by obtaining a light or medium weight fiber-filled bag. Slip the down bag inside the fiber-filled bag. The body moisture will filter all the way through the down material as it is warm from the protection of the outside fiber fill. When the moisture hits the synthetic fibers, it will continue on through leaving the down bag dry. If moisture does condense on the synthetic fibers, it is more easily dried than the down fibers.

Whether it is clothing or a sleeping bag, give some thought to your intended use, the terrain, temperatures, individual preferences, and other factors before making your purchases. There are also new materials, such as Gor-Tex, Bion II, and Thinsulate that you might investigate before purchasing a winter sleeping bag. Speaking of new materials, I have been very impressed with the Gore-Tex sleeping bag cover. This item is always included in my outdoor gear. This is a bivouac sack, made of Gor-Tex material, that slips over my sleeping bag. Condensation from body moisture can filter through this material, but rain or snow from the outside does not readily penetrate it. Several times during spring and fall excursions, I have crawled into my sleeping bag inside of a Gor-Tex cover with the stars shining brightly. At daylight, I would awake to discover several inches of snow on top of me, but I was dry inside the sleeping bag and the Gor-Tex cover. Besides letting body perspiration escape while deflecting outside moisture, this cover also adds five to ten degrees of warmth to your sleeping bag because of the added layer and trapped air.

I purchased my Gor-Tex bag cover from Early Winters, Ltdl, 110 Prefontaine Place South, Seattle, Washington 98104, but they are also available from other outdoor suppliers. You can obtain a recent issue of BackPacker Magazine, Outside Magazine, or similar publications and find advertisements from numerous manufacturers of these new products. Check the ads concerning jackets, gaiters, and other outdoor gear made from these new materials which you might be interested in obtaining as your finances and experiences justify. Whatever type of clothing and sleeping gear you settle for, remember that to stay dry is to stay warm. This might involve taking off or putting on a layer of clothing before starting to get too warm and perspiring or starting to cool off. It might involve slowing your pace, keeping an eye on approaching weather conditions, or other pre-planned actions. With proper clothing, gear and some experience, I am sure you will discover an uncrowded winter wonderland to be enjoyed with comfort and confidence.

CHAPTER 5
WINTER CAMP FOODS

Food for winter camps can vary greatly according to individual needs and tastes, but I prefer to keep meals simple. Summer campers can afford the time and energy expenditure for an occasional gourmet type meal, but with the shorter daylight hours and colder temperatures, it is advised that winter campers carefully plan menus ahead for conservation of time and fuel. By using nutritious and quickly prepared meals plus serving the foods and beverages hot, you are not putting out body calories to bring cold food or fluids up to your body temperature. Time and fuel may have priority during winter trips, so simple and fast prepared meals are an advantage.

There are numerous mercantile companies such as Mountain House, Richmoor, Chuck Wagon, and others that are marketing a wide variety of freeze-dried trail foods. Some of them have their own foil packets. All you need to do is add boiling water and wait five to ten minutes for the food to reconstitute and you have a hot, tasty meal. These main dishes are available in a wide variety such as chicken stew, beef and potatoes, beef and rice, chili, etc.

During the winter when everything is cold, you have to use special effort to keep these packets of food and water hot during the reconstitution period or they might end up tough and chewy. To do this, you can keep the packet submerged in hot water, or on hot rocks after adding the boiling water. If you set it aside on the snow while waiting for it to reconstitute, it will get cold in a hurry.

The Yurika retort dinners are fantastic but are heavier than some dehydrated foods. They are packaged with their own juices in laminated foil packets. Drop these packages into a billy-can or pot of hot water for several minutes to heat. You end up with a delicious hot meal with no cooking involved. The pouch can be opened and the food eaten without heating if necessary although this would be during emergency situations as hot food tastes better and does not take body calories to warm it up. I'm very impressed with the tastes and convenience of the Yurika foods which also includes a variety of quick mixes for fast campfire baking. These products are distributed by Yurika Foods Corporation of Birmingham, Michigan or check for a Yurika Foods distributor in your area. You should experiment and decide if you prefer the slightly heavier retort pouches or the lighter weight meals that require a bit of cooking time in camp.

The Yurika packets do not have to be kept frozen and yet retain a shelf life of several years. Winter camping allows you to utilize more different technics of meal preparation than does summer camping. For example, you can cook up batches of chili, stews, chicken and noodles, etc. at home before your trip. Put these cooked meals into seal-a-meal bags and into the freezer until needed. I prefer to make each seal-a-meal bag into an individual serving rather than make the bags contain portions for two, four, or whatever the number in the group. We can then pull out the required number of meals needed per trip. These water-tight bags can be dropped into a pot of boiling water in camp and you soon have a hot steaming meal. The individual serving bags thaw out much quicker than the larger portions do. These pre-cooked packages of food can be kept frozen during winter tours and used as needed. Breads, muffins, cinnamon rolls or fruitcakes can also be baked ahead and kept frozen until needed.

We like to cook up batches of scrambled eggs and sausage, scrambled eggs and ham, french toast, or other breakfast ingredients and package them in individual servings and freeze them in seal-a-meal bags. For breakfasts in camp, drop these packets into a can of boiling water to thaw and have a hot breakfast with no campfire cooking involved. If you don't mind carrying a few extra items such as honey, jelly, or syrup, you can also bake up pancakes, or waffles and package these in individual servings in seal-a-meal bags and freeze them in advance. When the packets are taken out of the hot water they taste like you had just taken them off of a hot grill. There is a powdered jelly that is handy to carry. You open up a packet and add enough water to make the correct consistency to spread over french toast or pancakes. Powdered syrup can also be cooked a little at a time over the campfire for those breakfast pancakes and waffles. Open a steaming package of scrambled eggs and ham to go with the pancakes, add a hot beverage, and top it off with hot instant Quaker Oats or Cream of Wheat which can also be obtained in individual packets. This gives you a good start for the day with a full fuel tank.

If you prefer not to eat this directly out of the plastic bags, you can use a paper plate and burn it when you are finished. This eliminates most of your dish washing chores. The stews, chicken and noodles, or meal of your choice used for the evening meals can be eaten directly from the bags which are then burned. It is no problem to keep these packets frozen while enroute during winter excursions by placing them on the outside edge of the backpack or sled exposed to the cold.

During the winter trips, always attempt to have some type of hot beverage such as instant coffee, tea, hot chocolate, hot jello, Tang, or other mixes with each meal. By drinking hot beverages instead of cold, it helps to put quick heat into your system as well as fluids which are essential. Personally, I do not care for coffee on winter trips because it has no nutritional value as compared to hot chocolate, or Tang which are giving you nutrition and fluids at the same time.

The noon meal can be much simpler, almost to the point of being just a "snack stop". Along with a beverage to insure the intake of body fluids, carry a few of the firmer backpack crackers and some cheese. Cheese has about two hundred calories per ounce and goes a long way. Mountain House has a very good freeze-dried tuna salad and a chicken salad in small packets. Just add cold water to the water-tight package, stir it with a stick and it goes well with cheese and crackers. Chuck Wagon and other suppliers also have a wide variety of these lunch type foods plus many items that require only 15 to 20 minutes of cooking for a nutritious evening meal. If you do not prefer to purchase the more expensive freeze-dried foods from the speciality outdoor stores, you can find many good items at your local grocery store. These would include Minute Rice, different types of instant potatoes, dried soup mixes, assortments of instant breakfast cereals, and dried fruits.

I do not carry a lot of cooking pots and pans on winter trips. Just a can for boiling water and a small cooking pot. I like a "billy can" for heating water. This is a number 10 can (or smaller) with a wire bail which can be set or hung over the hot coals of the fire. This way you always have hot water available for beverages, adding to the packets of food, or submerging the packets of food into for heating. If this "billy can" gets black or banged up after a trip or two, discard it and get another juice or lard can, punch a small hole in each side at the top of the can to fasten a wire bail on. This can works just as well as some fancy utensil purchased in a store and is a lot more economical. A breadwrapper or a plastic bag can be slipped over the blackened can after each use to keep your pack clean. Food packets, cups, or other items can be stuffed into the "billy can" to save space when packing.

If metal cups and plates are placed by the fire they will get so hot that they will burn your lips or you cannot even hold onto them. If they are placed on the snow they get cold and the food quickly cools. The Sierra Cup with the cooling metal rim can be serviceable in the winter. Personally I like rigid plastic cups and bowls providing they can withstand boiling water. Do not leave them too close to heat or the fire because they will melt. You will remember this precaution after a slip-up or two. These plastic utensils weigh very little and are easy to clean. In fact, I use few dishes in a winter camp. When I am finished eating, I can dip a bit of hot water from the "billy can" into the bowl or cup, swish it around a bit, and wipe it out with a paper towel which can then be burned. You might prefer to carry an extra cooking spoon or two with each person carrying their own knife, fork, and spoon. If you would like to keep your load as light as possible during winter tours, a spoon is all that is needed for about anything you could eat in a winter camp. At times, when neglecting to take my spoon, I have carved a rough spoon from a flattened stick and gotten by very nicely.

I do not carry soap with me on winter tours because it is not necessary, especially if each person wipes out their own bowl and spoon or washes them in the snow. This way, you do not need an extra pot of hot water to rinse the soap off of the washed utensils or risk a case of diarrhea from soap left on the tableware. During cold weather, bacteria will not grow on any left-over tiny particles of food like it does during the summer. By trimming the bulk and weight of cooking and eating utensils on winter tours, this allows more space for the extra cold weather clothes and gloves.

I might also be carrying weight and bulk with pre-cooked breads, doughnuts, or fruitcakes because there are less hours of daylight around camp to do this type of cooking. While on winter tours it is not as convenient or comfortable to sit around a campfire while attempting to bake in a reflector oven as during summer tours. When carrying frozen bread, muffins, etc. on winter tours, you can take out the desired quantity ahead of time, put it in a plastic bag, and place it in an inside pocket in your shirt or coat. Body heat will thaw it out for the next meal. If using bread inside plastic wrappings, do not attempt to thaw it by the fire because the plastic can melt and be difficult to separate from the bread slices. If frozen, put the bread slices on a forked stick over the coals and it will soon be thawed out and toasted.

You can prepare ahead of time various types of snack foods for winter excursions. Assortments of these mixtures are ground and mixed together into bars called pemmican, of which there are numerous variations. Each individual or group seems to have their favorite concoction of dried fruit, ground grains, chocolate, peanut butter, honey, nuts, and other ingredients. A simple favorite of mine is to purchase a pound each of raisins, peanuts, and chocolate chips. Dump them together in a bowl, mix them up and store them in smaller portions in plastic bags. These bags can freeze or thaw with no harm until utilized from the pack or survival kit as a quick and easy snack.

You should carry some type of canteen or water flask with you during the winter because the intake of fluids is essential for body efficiency and warmth. When traveling over the snow, there is often no running water visible and you usually do not stop to melt snow for drinking water while traveling, thus the importance of the canteen with fluids. If weight is not a big problem, some winter campers like to carry a small thermos jug. This can be filled with hot chocolate or hot soup. At mid-morning you can stop for a "mug-up" consisting of several sips from the hot thermos along with some fruitcake, cookies, or doughnuts that has been carried in an inside pocket so that they can be thawed. You can finish consuming the contents of the thermos bottle with a similar stop in mid-afternoon. This pause with a hot drink and a snack helps to warm you up, let your body rest a bit, and give you a boost of energy to continue on. The thermos bottle can be washed out in camp each evening and if filled with water, it will not freeze in a tent or lean-to. That way you can have a drink during the night if desired. If sleeping in a properly constructed snowcave, a plain canteen of water will not freeze like it would in a tent or lean-to. Before you leave camp the next morning, again fill the thermos with a hot beverage or soup for the mid-day stops.

The foods for winter camping can vary greatly in content and type depending on your desire for economy, pre-preparation, weight, bulk, and other factors. To help promote body energy and heat during winter tours, include more fat type foods such as cheeses, sausage, peanut butter, and nuts; than on summer excursions. Fat type foods are also slower digesting than sugar and carbohydrate foods. By mid-morning, when the sugar and starch foods are about used up, the fat foods are still digesting and going into your system for body energy and heat.

When exerting energy to travel over the snow plus the need to produce body heat in a cold atmosphere, you will require a lot more calories than on summer excursions. So include extra foods for this double need. Remember that fat type foods generally contain more calories per ounce than starch and carbohydrate foods such as breads and noodles. Experience and individual preferences will soon allow you to plan and better prepare your gear and food for excursions into remote areas without being burdened down with unneeded items. Start with short tours and increase them as you gain experience and confidence.

CHAPTER 6
CAMP TOOLS AND GEAR

In an emergency situation, you can improvise various types of tools to dig a snowcave. This could include your hunting knife, a cooking kettle, a stick or tree limb that has been chopped or whittled flat for digging or cutting into the snow, a hub cap from your car or some other object you can find for a tool. If you are not carrying a tent or other shelter on your winter excursions and are intending to utilize nature's materials for nightly protection, make sure your pack includes some type of shovel or other implement for use in snowcave construction rather than having to improvise.

When shopping in various outdoor supply catalogs, the inexperienced person can run into problems by ordering the wrong items. I have seen many types of small and light-weight shovels advertised in these circulars. They might work satisfactorily for a few minutes in your back-yard, but when you are on a five or six day cross-country tour where you might have to construct a shelter during a raging blizzard, you want a dependable tool. Sure, those catalogs illustrate a nice aluminum shovel that does not take much space and will add little weight to your pack. Some of them even come apart so as to fit in a small space for easy carrying. You took it out in the backyard and it worked fine in the new fallen snow. But what about some of the hard, wind packed drifts in the mountains, or the icy crusts where the sun has melted the snow a bit and then it has frozen again? The sun is going down with a storm approaching, and you are in a hurry to get the night's shelter completed, and exert extra pressure in a solid snowdrift. Bingo! The aluminum folds and the shovel breaks off right at the handle.

When you are ordering or gathering equipment for your winter excursions, keep situations such as the above in mind. Some of the light-weight shovels are constructed and reinforced in such a manner as to be quite durable and useful while others might look nice, but will not stand an extra strain. Of course, this can also vary with the mood and experience of an individual. I can take one of the more delicate shovels and carefully dig a snowcave with it. Past experience tells me how to handle it and how much pressure to apply. But I can hand that same shovel to an inexperienced snow-caver on one of my cross-country tours and it will generally be broken in five minutes.

It does not take a large shovel to dig a snowcave. My choice is a small, round steel blade about eight inches long by six inches wide with a solid wooden handle about eighteen inches long. It is rigid enough that I can use it to chop through crusted ice or snow, yet small enough to fit into my pack. It weighs a bit more than an aluminum shovel does, but it is very sturdy. I have dug dozens of snowcaves with this small steel shovel. I have also loaned it out to other participants, and it is still in excellent condition. I used to purchase this shovel from Army Surplus for 98 cents many years ago, but I have not recently found any like it. Most surplus catalogs now list the folding trench shovels, but this is much too heavy to carry in a pack and it is also awkward to use with it's straight blade design.

An all plastic shovel manufactured by Life-Link of Jackson Hole, Wyoming, has also proven to be very satisfactory for winter tours. It has a square blade about eight inches across, is quite durable, and is light in weight. The handle and blade come apart so that it can be easily carried in a pack. On some models, the handle can be extended to make it longer for digging purposes.

When I take a group of scouts or youths out snowcaving to a location that is not far from the end of the road, I often take several large grain scoop shovels along, especially when using a snowmobile and sled to carry the equipment. You can move a lot of snow in a short period of time with these large shovels. However, they are not practical for finishing the inside of a snowcave or carrying in a pack because of the large blade and the longer heavy handle.

There are many styles and types of shovels available that can be utilized for snowcave construction. Some are sturdy and can be relied on for extended tours while others cannot. Some are heavier than others, so you have to experiment and then decide for yourself which type and model is best suited for your needs. When you are traveling via vehicle or sled, weight is not as critical as when you are carrying it in a pack on your back. When you are traveling with a group, the gear can be divided up with one person taking the shovel while another person takes the cooking utensils or food. Keep the packs light by not having each individual duplicate unnecessary items.

Some shovels have extendable handles and fit into a backpack. I (author) prefer the smaller solid handled shovel.

If you have only one shovel in a group of several people, it is a simple matter to keep that shovel working by trading back and forth. One person can shovel rather rapidly for short periods of time, then that person can rest while another individual shovels for a while. It is a bit slower to complete a snowcave in this manner than it is with several shovels, but your packs are lighter.

Some folks like a machete with an 18 inch blade for slicing the snow into chunks and then tossing them out of the snow cave by hand. On occasions the snow is not of a suitable texture for cutting into large pieces for pushing out of the excavation, but crumbles when moved. The machete is also good for making blocks of snow for igloo building. Try out the flat aluminum sheet described earlier in Chapter 3 on igloo building. It weighs very little and easily fits into a pack, yet will move a lot of snow, but is more suited for cutting igloo blocks rather than digging a snowcave.

After some field experience, do some thinking and come up with some of your own improvised tools for snowcave and igloo construction depending on the snow texture and conditions in your area. Keep in mind that you will not always find the same texture of snow. Sometimes, it might be rather solid in windblown drifts where you can cut or break off the snow in chunks and toss it out of the snowcave. At other times, you might find the snow granulated and will have to move it out one shovelful at a time. In this situation, the larger shovel blade is more efficient while in the more compact snow, a small rounded shovel blade does a very nice job of cutting or breaking off the larger chunks of snow to be tossed out. Check the advantages and disadvantages of size and weight in the pack, durability, construction time, and other considerations most likely to be encountered in your area and go from there.

For extended cross-country trips, you will need a backpack of some type to carry your food, sleeping bag, tools, and other gear. There are many different models and types of packs available on the market, but DO NOT end up with a "cheapie". When finances are a problem, the economical type packs are okay for the beginning youth or the occasional overnight packer, but I cannot recommend them for trips into remote areas. They have a habit of coming unsewn or ripping in the middle of nowhere. The sturdier made packs will generally stand up much better under strain or long use conditions.

I personally prefer the Coleman Peak I backpack for several reasons. The frame is made of a durable plastic material that is of a stiffer texture at the bottom, a medium stiffness in the middle, and a more pliable material towards the top. This flexibility of the frame eliminates any sudden jars or strains due to stumbling, bending over, etc. There are several dozen slots in this plastic frame which allow quick and easy adjustments considering the user's size, heavy or light clothing, or type of load being carried. It also has numerous exterior or side pockets for carrying snacks, gloves, film or other small items that are frequently needed. These pockets eliminate undoing and searching through the entire pack each time something is needed.

Many other brands of good quality backpacks are available from various sources. You can obtain recent issues of outdoor magazines, such as Backpacker and Outside Magazine, and write to the advertisers for more details and information on items you are interested in purchasing.

If you are skiing with a backpack, problems can be created from the swaying load. You might prefer the internal frame pack which rides snug against your body. The internal frame pack is narrower than the external frame pack which gives more freedom for your swinging arms when using ski poles. During winter use, perspiration on your back with the snug pack is generally no problem as it might be during the warmer summer months.

It is suggested that you develop a system of stowing gear in your backpack. Stow items such as sleeping bags, pads, and cooking utensils that are not needed during the day in the main pack. Put the layer clothing, days' lunch, and other items you might need during the day towards the top of the pack. Utilize the smaller outside pockets for the frequently used smaller items. Camera filters and film can go in one pocket. Sun screen, Chapstick, Bandaids, and personal hygiene supplies in another pocket with the midday snacks or fire starter in another compartment. Get in the habit of keeping related items together and always in the same place in the pack. You will then know exactly where to reach when that item is needed.

There is a limit to what you can comfortably carry on your back, especially when traveling on skis or snowshoes through various textures of snow. For example, the snow can be crusted on open windblown slopes or it can be fluffy in the tree-shaded areas. The weight of the pack on your back makes it much more difficult to maintain balance unless you are an experienced skier. If on snowshoes, the added weight of the pack makes you work harder for each step. If you do happen to lose your balance and fall forward into the deep fluffy snow, it is difficult to regain your footing with the weighted pack on your back. It is usually easier to slip out of the pack, get up and brush off the snow, then grasp the pack and slip it on again.

These factors can vary according to the experience and dexterity of each individual. For winter travel, ladies should attempt to keep their packs at about thirty pounds or less while the men can carry more weight. This can vary with the size and experience of each participant. For example, I have a lady guide of medium size that can carry a fifty pound pack all day while cross-country skiing. At the same time, I have seen some two hundred pound men have problems with a thirty pound pack.

If the backpack load and equipment is too heavy for a backpack, then try the small fiberglass sleds. Do not purchase one of those kid's toy sleds from a local shop, tie a rope onto it, and take off on skis or snowshoes. On the downhill trails, there is nothing to stop it from sliding into you and knocking you off of your feet. You can get specially made sleds for cross-country tours, recreational and work uses. They are made of durable fiberglass and are about eighteen inches wide and four to five feet long. The entire sled weighs eight pounds or less. Larger sizes are available, but are generally not necessary. They come equipped with long telescoping shafts fastened to a padded, adjustable waistbelt with a quick release buckle. This gives you constant control of the sled on flat terrain, uphill, or downhill.

I obtained my sled from MountainSmith, 12790 W. 6th. Place, Golden, Colorado, 80401, and am very satisfied with it. Besides using it on various cross-country tours over the Continental Divide, I have used it on my winter traplines in the mountains where the nearest human track is ten to fifteen miles away. I often carry over one hundred pounds in this sled behind me with little extra effort depending on the terrain. The loaded sled with the smooth, flat bottom slides over the snow surface very easily and the small runners help maintain it on sidehills. Some of these sleds have a built-in cover that can be lashed shut after stowing the gear. I prefer the sleds with several small rings on each side. Take an 8' X 10' tarp or other appropriate size, fold it in half and lay it on the sled. Put your backpack, shovel, and other gear on this cover which is then folded over and lashed secure with a nylon rope or cord. Place the heavier items on the bottom with lighter objects toward the top for better balance. When making camps, the tarp and the extra cordage can be utilized for making a wind shelter or keeping other gear dry. By checking around, you can probably locate other manufacturers of similar type sleds and shafts in other sections of the country.

It is easier to transport heavy loads over the snow on a light fiberglass sled rather than in a backpack.

If you are going out for short daily excursions where you do not need to carry extra gear, such as sleeping bags, or shovels, you might prefer to utilize some type of small day-pack rather than having your pockets stuffed full. There is a wide choice of small daypacks available for carrying lunches, extra gloves, cameras, and smaller items. The shoulder straps do have a tendency to restrict movement under clothing which retain perspiration dampness. There are some of the daypacks that have waist bands like a backpack, which helps to take some shoulder pressure away. I prefer the fanny pack which straps around the waist. It also rides on your hips which is the strongest part of your body. Then your shoulders are free from any restrictions. You can carry this light fanny pack on your hips all day and not know you have it on but, when you need it, it is there. A disadvantage of the fanny pack is that it does not have ample size to carry extra garments in order to utilize the layer system of clothing if the weather changes. Garments can be lashed on top of the fanny pack, but they are bulky.

The mode of travel over the snow is again an individual discretion. Some folks prefer skis, but if you are inexperienced and in brushy terrain, snowshoes are recommended. If you are not familiar with the use of either one, you can generally rent them from outdoor shops and try the various sizes and styles before making your choice of purchase. Each method of over-the-snow travel has advantages and disadvantages. In average terrain, an experienced skier probably puts less physical effort into each mile traversed than does the snowshoer. The snowshoer usually gets better results on the uphill trails while the skier gains on the downhill terrain. Much of the time, the skier can glide along without much effort while the snowshoer has to pick 'em up and set 'em down with each step. Folks that are new to winter recreation can learn to snowshoe much easier than they can learn to ski and they are more economical.

If you prefer to use snowshoes, there are many options to consider. The wider bearpaw designs are good for brushy terrain, but are a bit more difficult to use than the narrower trail snowshoes. The rawhide laced webs take more care and an occasional varnishing, whereas the neopreme laced snowshoes require less care. A large person needs a larger snowshoe for floatation on the snow, while a small person can use a smaller snowshoe. Snow texture is also a factor on snowshoe size. Some snowshoes will slide on crusted snow or sidehills while others, such as the Sherpa models, have flanges underneath that bite into the snow to keep you from sliding. The upturned toes as on the Yukon Trail models are a great advantage to sliding forward in fluffy snow.

The bindings that fasten the snowshoes to your feet can vary greatly. Some bindings are a sandal type affair while others are a series of straps. Some bindings are made from rubber that gives with each step rather than drawing tight across the back of your ankle which can cause a tender spot after several consecutive days of snowshoeing. As with the other equipment, you make your own choice after some experimenting depending on your weight, average type of load, snow conditions, terrain, and other factors. If you are a confirmed ski addict, you would probably not consider using snowshoes.

The author shows some of the different types of snowshoes available

Snowshoe bindings should have a hinge action on the snowshoe, but must be snug for control. Leather sandal binding on Sherpa showshoe and rubber binding on a modified trail snowshoe.

Your companion or instructor might advise or suggest this or that type of equipment. These suggestions are not to be taken lightly because that person has probably utilized different types of equipment on previous occasions and has come to a conclusion of what is the best for him or her. These suggestions can perhaps save you considerable time and expense in testing assorted gear. However, remember to enter your winter camping with an open mind. You might come across something more practical for your use than those techniques that were pointed out to you.

CHAPTER 7
ON BEING PREPARED

Aside from your basic equipment, there are numerous smaller items that can greatly add to your comfort and well being, not only for basic winter camping, but during unforeseen situations. Outdoor campers should carry a survival kit with them AT ALL TIMES. This packet should contain a fire-starting kit and some type of body shelter. You can add other items to commensurate with your needs and choice. This kit should remain small and compact. If you get it too large, you will have a tendency to leave it in camp or in the vehicle especially when you are not planning on going very far or do not anticipate needing it. Most emergencies happen when you do not expect them.

Perhaps you intended to only go over the ridge a short distance and be back soon. But some tracks, plants, or scenery captured your interest and you proceeded farther than expected. All of a sudden you look up and the sun is down, a fog or a storm is moving in, or some other unexpected problem arises. You turn to go back, but the landmarks do not appear to be the same. You had neglected to occasionally look behind to check the landmarks from that angle for your return route. You become disoriented as darkness approaches and your survival kit is back at the vehicle or camp. This is very serious, especially during the winter when temperatures become very cold and wind can erase your tracks in the snow so you cannot backtrack.

Another example of an unexpected problem arising would be if you were to slip on the rocks, sprain an ankle or break a leg and be unable to return without assistance. You would be required to spend the night where you were injured. Some source of heat and body shelter would be of a great benefit in this type of a situation. Develop the habit of carrying a survival kit with you on all outdoor excursions---short or long. With a survival kit containing a dependable fire starter, you could easily get a warming fire going. A piece of plastic or a wind repellent jacket will help you stay dry and give you protection from the wind.

There are numerous types of firestarters that can be purchased. You should do some experimenting and utilize what is best for you. I do not depend on regular matches for emergency situations. When carrying matches in an inside pocket, they may become slightly damp from perspiration or moisture and be difficult to light. While attempting to strike one of these matches, the head sluffs off or, if you are successful in getting it to burst into a flame, that flame fizzels out before reaching the tinder. If the tinder is slightly damp, as it often is during a storm, that flame of twenty seconds or less is not long enough to get the tinder started, especially if a cold wind is blowing and you are shivering from the damp and cold. In this type of a situation, a small candle or piece of candle carried in your survival kit can be a lifesaver. By lighting the candle and protecting it from the wind, you can extend that twenty second match flame into a small several hour flame. Put it under the damp tinder which will soon dry out from the constant flame and start to burn. Then you can add larger pieces of tinder or wood. After you get your fire going, reach under the fire with a small stick and flip the candle out to save it for future use.

There are numerous types of fire-starters available in some of the outdoor stores. Be sure that you test them and are familiar with it before you stuff it into your survival kit. During a damp, cold, or windy situation is not the time to learn how to use it.

I personally like to depend on the metal match, but you should know how to use it before going into a remote area. The metal match is a small shaft of metal compounded from several different metals and when it is struck or scraped with a fast movement from a sharp hard object, it will emit numerous sparks. These sparks can be directed to a piece of tissue paper or dry tinder and with the proper technique it will start a flame. However, with a piece of tissue paper or tinder, you can have the same problem as you do with a damp match. That is maintaining the heat long enough to get a substantial flame started. So I use the metal match in conjunction with a chemical compound. This can be a tube of fire ribbon, a small can of sterno, or other firestarting materials sold in most of the outdoor stores. These materials can be lit in the tube or can to start a fire to warm your hands. A tube of what is commonly called "airplane glue" also does a good job. It has an ether base and a spark will readily ignite it. However, you need be careful with such a substance that has an ether base. If you try to light the tube, the flame could crawl down an air pocket in the tube and explode. Therefore, whenever you are working with an ether base compound spread some of it onto a stick, screw the lid back on, and set the tube aside. Grasp the metal match and scratch a spark onto the dab of glue and presto, you have a hot flame burning. Then add the tinder and wood that is needed to build a larger fire.

A cigarette lighter makes an excellent fire-starter if you make sure it has fluid in it. They are inexpensive and should be purchased, tested, and packed away. A couple of flashlight batteries making contact with a twisted piece of fine #0000 steel wool can be used to start a fire. You can make your own tinder for starting fires by soaking strips of cardboard in melted wax. I sometimes use a bow and drill to get a fire started, but this is not recommended without a lot of practice. It can be useful in emergency situations, but be sure that you are familiar with this process before attempting to build a fire during adverse weather conditions. Whatever method you use to get a fire started, be sure you have ample tinder and fuel wood gathered before you attempt to start it. I have seen individuals start a fire with a handful of tinder available. When they get the tinder burning, they then rush into the brush for more small sticks to put on the fire. There is often not enough heat left in the original tinder to get the new pieces started and the flame fizzles out. So have everything ready before you attempt to start the fire.

After you get that fire started, you can reflect more heat back toward you by piling up rocks or logs on the opposite side of the fire. By taking advantage of a rock outcropping or a lean-to you have constructed from available material, the heat from the fire reflects against you as well as bouncing off of the lean-to or rock and down onto your backside. With a small warming fire by a lean-to on one side and a reflector on the opposite side, you can get through a cold winter night in comparative comfort.

I have spent many comfortable nights during winter blizzards by lying or sitting under a lean-to next to a small fire. Gather a supply of firewood before dark and have it stacked close to where you are lying or sitting. If you doze off and become chilly as the fire begins to burn down, you will wake up and can then reach over and put another chunk or two of wood on the fire without getting up. By piling brush, boughs, or snow up on each end of the improvised shelter, it deflects any side breezes and helps to retain some of the reflected heat from the fire. Utilize leaves, pine needles, or boughs under you to help retain body heat.

In addition to having a dependable fire starter in your survival kit, you should have some type of body shelter. You could use a compact nylon jacket to serve as a windbreaker and to help shed moisture during a storm. A sheet of flexible plastic is very compact in a survival hit and makes a good windbreaker. A sturdy garbage bag or two can serve for this purpose. Cut a hole in the bottom center of the bag, just large enough to slip your head through, then slip the bag down over your body to deflect wind and moisture.

There are numerous other items that can be carried in your survival kit depending on your training and background. As mentioned elsewhere, remember to keep that kit compact so it can be carried with you at all times when you are away from camp or vehicle. I always carry a small roll of stout nylon cord. It can be used for many different purposes, such as a snare to catch a snowshoe rabbit for a meal, lashing a lean-to together, emergency snowshoe binding, and other uses.

Speaking of snares, I like to carry a small commercial wire snare with me on cross-country tours. It is easy and quick to set up and it is more dependable than a primitive type snare made from a piece of cordage. In much of the high mountain country, a snare can be set in the snowshoe rabbit trails for a chance of capturing a rabbit to supplement your food supply during emergency situations.

Your kit should contain food items, such as candy bars, raisins, nuts, jerky, and other good nutritious items of your liking. Do not overload your small kit with food items rather than the essentials such as the body shelter and fire starter. You can go many days without food if you pace yourself to conserve body energy and heat, but you do need fluids in order to maintain efficiency.

Carry a couple pieces of folded aluminum foil in your survival kit. This can be shaped into a utensil to melt snow or ice over the fire so that you can have drinking water. DO NOT eat ice or snow as it takes body calories to heat it up to body temperature. With the use of the foil, you can obtain warm water which can be supplemented with powdered cocoa or bouillon cubes from your small kit. DO NOT neglect the intake of fluids during cold weather even though you do not feel thirsty. You are losing body moisture through respiration, urination and possibly perspiration under heavier garments. Body fluids are important for circulation to help ward off hypothermia.

I always carry a piece of sinew in my survival kit. This is a length of tendon from the backbone or leg of a carcass, such as a deer or calf. When needed, it can be softened with moisture, tapped a few times between two small rocks or a knife handle, and then stripped into dozens of fine strands. These small strands can be soaked in water or held in your mouth for several minutes so they become pliable. Then the sinew can be used for a number of things, such as repairing clothes, wrapping around split knife handles, fixing broken glasses, etc. It is not necessary to tie any knots in the sinew because by simply pressing the soaked strands together on top of each other, the natural mucus will hold it together as it dries. I have repaired many items on the trail with sinew and have had good results. However, for clothing repair you might prefer to carry a needle and thread in your survival kit.

A small flashlight with extra batteries and bulb should be added to your kit along with a few first aid items. Do not neglect to carry a compass and map with you, but be sure that you know how to use them. As can be seen, you should make up your survival kit according to your needs and desires and the time of year and terrain you are going into. I have never been satisfied with the various commercial survival kits found in the stores. They usually contain items of little use for my terrain or season of the year. I prefer to carry only those items that are useful to me. My survival kit ALWAYS contains a pair of rain chaps. These are leggings made from moisture resistant nylon type material. They can be slipped up your leg over the boots or shoes and tied to your belt. They help to keep the wind from carrying away body calories on cold days, keep your legs dry during storms, and are very effective in keeping your legs dry while digging snowcaves. The chaps roll into a bundle about four inches long by one and one-half inches wide. Consequently, rain chaps take very little space in a survival kit and can be extremely useful in staying dry or maintaining body heat. Rain pants can serve the same purpose, but perspiration cannot readily escape which can create problems. With the rain chaps fitting loosely over the trouser legs and being open at both ends, air can move through them with a chimney effect as you move and carry some of the perspiration away.

You will need something in which to carry your survival kit. If you are on a short excursion and are not using a large backpack or sled, you will then want something small. Some folks utilize the small day pack to carry their survival kits. As mentioned, the shoulder straps may constrict upper body movement and do not allow perspiration to readily dissipate. Therefore I like a fanny pack for carrying my survival kit. This is a small pouch with waist straps which puts the weight on your hips and leaves the shoulders free.

Along with the survival kit and knowing how to utilize it, is your own positive mental attitude of, "I can do it - I can do it". This can be a valuable asset, especially during emergency situations. If you allow panic and psychological stress to take over, you can use up to seven times as much energy as when you stay calm and pace yourself. During a stress situation, the adrenal glands can drain body energy rapidly if not controlled via PMA. This is easier said than done and where the experiences of coping with storms, disorientation, darkness, or related problems, can always be of future value if remembered and applied. Cold weather and winter storms are no great problem if you are prepared and know how to cope. In fact, I enjoy matching wits and skills with winter storm situations as it gives me a sense of accomplishment. It also helps a person to realize that one does not work against Mother Nature, but with her.

For example, DO NOT take chances with snow avalanches. During winter tours, watch for old avalanche paths where the snow and ice has slid down the mountain in past winters. It can happen again! If forced to cross these areas, attempt to stay high and above any fracture lines. When this is not feasible, you can sometimes make your route along the bottom on the far side of the slide area. If there are two or more persons traveling together, move across these danger areas one person at a time. If an avalanche does occur, the person waiting at a safety zone has a better chance of seeing where the victim might be caught in the slide and then be able to assist him or her after the slide has stopped.

If your winter excursions will be taking you into areas with an avalanche potential, you might desire to check into avalanche probes and avalanche cords. An avalanche probe is a thin rod ten to twelve feet long that can be poked into the snow in a systematic pattern when attempting to locate a victim buried in the snow. Some are made of light aluminum tubing with a cable through the center so it can be pulled apart and folded in order to place it in a backpack. An avalanche cord is a bright colored string attached around the waist of a person traversing a potential avalanche area. If buried in a snow slide, the thirty or forty foot cord with yards or feet marked on it are apt to float towards the surface in the tumbling snow. The rescuers can sometimes readily locate this red or orange cord on top of the slide area and follow it to the buried victim.

Most avalanches occur during or shortly after winter storms, during temperature changes, when sustained winds move snow onto leeward slopes, and other factors. If you intend to spend considerable time in potential avalanche terrain, it is suggested that you attend an avalanche school, or at least read some of the books and literature on avalanches. This avalanche literature can be obtained from the Forest Service offices in snow country.

This chapter is not being written to discourage you from enjoying the outdoors during the winter. Instead, I am attempting to alert you to be prepared for unforeseen situations, both mentally and physically, and to be able to anticipate the weather moods and terrain of the great outdoors. With some physical conditioning, a positive mental attitude, some experience and proper equipment, it can truly be a winter wonderland for you.

CHAPTER 8
SUGGESTIONS FOR ADDED COMFORT AND ENJOYMENT

By the time you have reached this last chapter, you have read about various methods of constructing snowcaves, igloos, and other types of body shelters. You have also read suggestions on winter foods, clothing, and other miscellaneous gear. Much of this information sounded simple to put into actual use. If you are just starting winter camping and everything does not end up as pictured in this book, do not despair. It took me numerous attempts to develop the most efficient methods for the desired results. Each time I go out snowcaving with a group, I still stumble onto new helpful tools and methods for my projects. Who knows, maybe ten years from now my snowcave construction methods may be considerably different than those I now use.

I purposely did not devote a lot of space to the use and types of snowshoes, skis, sleds, clothing, avalanches, and other related winter subjects because entire books are written on each of these subjects. I attempted to discuss each of these topics enough to illustrate that there is a wide variety of choices and approaches to any and all of these subjects. My main purpose is to acquaint you with the insulating qualities of snow and to instill in you a desire to utilize that abundant material for both added enjoyment of the outdoors and for use in emergency situations.

As you begin to utilize the winter wonderland, you can obtain literature on the above subjects to further increase your knowledge and skills of the outdoors.

Before concluding, I would like to suggest several precautions that could be important while you are in the process of learning on your first snowcaving excursions. To start with, DO NOT neglect your eyes in the white winter environments. Snow blindness can strike you and it can be corrected only by keeping your eyes covered for several days. As a precaution against snow blindness, always wear dark glasses or goggles in snow environments. Even on hazy or cloudy days light reflects from both the snow and the clouds and it can cause temporary blindness.

Keep in mind if you are wearing sun glasses rather than goggles, bright light can come into your eyes from the side. During prolonged periods in the winter brightness, you can improvise by putting tape (from your first aid kit) on the sides of your sun glasses to help keep out that bright sidelight. It is possible to purchase sun glasses which are made with side pieces specifically to prevent this problem. If you should lose or forget your dark glasses or goggles you can improvise by cutting narrow slots in a piece of bark or wood that is trimmed to fit your face. The narrow slots eliminate a lot of bright light coming into your eyes. A piece of shirt tail or a strip of cloth will serve the same purpose as would a bandanna or scarf.

As with all phases of outdoor activity, use common sense and judgement for your well being. You are your own destiny and the mountains don't care.

Be sure to take extra gloves or mittens on the snowcaving tours. Until you have checked out and obtained the desired technics and equipment, you are apt to end up with wet gloves or mittens when digging a snowcave. The friction of your hand working on the shovel handle or other type of tool creates some heat. This will melt some of the snow you come in contact with, and this melted snow can in turn freeze into ice which makes for cold hands. After some experience of digging in the snow, you will learn not to get your gloves or mittens damp. But there are times when lifting the snow blocks for the cave wall that your hands have to come in contact with the snow. The snow will brush or fall off during cold weather, but if not careful during moderate and warm temperatures, you can become damp very quickly. Whatever the case may be, have extra gloves along for using while the others are drying. You might consider carrying a pair of rubber gloves for use while digging your snowcave. A light polypropylene glove or liner worn under a sturdy dishwashing rubber glove has been very satisfactory for some folks.

You need to be alert and take precautions against frozen fingers and toes. If your fingers, toes, ears, nose or cheeks begin to sting from the cold, immediately heed those body indicators. If you should decide to ignore these stinging sensations and continue to complete this or that camp chore, you could end up in serious trouble. As you continue the task, the stinging sensations may seem to fade away. At this time, you might think there is no problem because you no longer feel the cold. Wrong! The nerve endings in those parts are being frozen, hence the reason you do not feel them anymore. If you continue on with the chore without stopping to warm the cold parts, you can create even more frostbite damage without knowing it until later. Remember, give immediate attention to those body indicators before severe damage occurs!

It is sometimes necessary to kneel while constructing your snow shelter. An old sack, if available, or just an evergreen bough under your knees can help to keep them dry as you work. If rain chaps, as described earlier, are in your survival kit, utilize them to help stay dry and warmer.

Another hazard to be aware of during winter excursions is a white-out condition. Unless you have experienced a severe winter white-out, it is difficult to imagine the results. You do not necessarily need a snow storm to create this condition. The sky can be blue without a cloud in sight. A wind can come racing across the mountain and pick fine particles of snow up off the ground and create an apparent fog. A snow storm, along with a high wind, has great potential for white-outs. When the sky is cloudy, there is sometimes an atmospheric heaviness that makes it very difficult to tell where the earth stops and the sky begins. This situation makes it nearly impossible to see the ups and downs of the terrain until your skis or snowshoes come in contact with the various terrain changes. Then you don't see it, you feel it!

I recall one incident when I was escorting a group of college students for education credits on a snowcaving tour. With these groups where it is often a first-time camping experience for some of them, I always attempt to have a back-up shelter available for safety reasons. Some of them do not have the proper clothing for winter trips (cotton blue jeans rather than wool) and do not realize that to stay dry is to stay warm. If they become damp or cold while constructing their sleeping caves or igloos, I have a plastic shelter twenty feet long and twelve feet wide set up in nearby trees where they can warm up. This temporary shelter has benches and tables inside with a small woodburning stove in one end and a propane heater in the other end.

On this type of tour, the participants ski or snowshoe to the campsite several miles from the end of the snowplowed road. The students construct snowcaves or igloos in the deep snow drifts near the temporary shelter. I can circulate back and forth in between individuals to assist or instruct where needed during the learning process. After completing their snowcaves the participants can filter into the heated shelter for warming, eating, evening lectures and demonstrations on other survival skills. The damp mittens, socks and other articles are hung up on suspended ropes to dry. At the end of each evening, they go out into the darkness to sleep in the snowcaves or igloos constructed earlier. The first evening seems to be difficult for them to go out in the dark and crawl in that hole in the snow as they are sure the night in that snow shelter is going to be a very cold one. It is interesting to have them filter into the shelter the next morning and state, "Hey, that wasn't bad! I didn't get cold at all! I have seen the time my college dorm was colder than that snowcave". Perhaps an exaggeration on their part, but the warmth of the snowcaves had made an impression on them.

One of my guides has a hot breakfast waiting for them in the shelter each morning. In this manner, they learn the basic principals of winter camping. If they make a mistake in planning such as type of clothes or perspiration wetness while working, it is not critical because of the warm back-up shelter. After a tour of this type, they are then better prepared for actual future cross-country excursions where they will carry everything in a pack with all of the cooking and eating done in or near the improvised snow shelters.

I recall one tour when it had been snowing hard for three days with a wind creating a very severe white-out. After breakfast on the third morning we were sitting around the table in the shelter discussing survival technics. One of the ladies left to go to the latrine about forty yards away in the trees. After a while, someone remarked that she had not come back yet. My lady guide bundled up in insulated garments and went outside to the latrine to check on this individual. She came back and reported the lady had disappeared!

We immediately put on heavier outer garments to start a search. Yelling would do no good because the howling wind and falling snow absorbed all of the sound. All of a sudden the wind slacked and the sifting snow lifted for a brief minute. We glimpsed the missing lady floundering in the snow in an open park about one hundred yards away. Then the white-out closed in on us again. We took a compass bearing, worked our way through the blank whiteness of falling snow and fog, located the lady and brought her back to camp. This incident made a lasting impression on the participants whom were then very cautious when venturing out into the storm. The experience of not being able to see a large green plastic shelter, in the whiteness, only a few yards away illustrated to them how severe a white-out could be. The trail to the latrine was marked for night use. This can show you how meaningless directions can be during a white-out. It is very frightening if you are by yourself and have never experienced anything like it before. By going with instructed groups you can learn how to handle yourself in these conditions and then feel more secure if stranded in a solo situation.

On another occasion, most of the participants were into the base campsite and I was coming in with the last load of grub, sleeping bags, and other gear tied on a long sled behind the snowmobile. One of the participants had gotten behind the group on skis and I invited him to ride with me on the snowmobile back to the campsite. About two hundred yards from the campsite, which was on the other side of an open park, the loaded snowmobile would not negotiate through the deep, fluffy snow as I started to go up a slight hill. I unloaded part of the load from the sled and told my passenger to wait by it while I broke a trail into camp with the lightened sled. Then I would return in a few minutes to get him and the remaining gear. As I drove away from my passenger the storm slacked off a bit, but while I unloaded the toboggan at the campsite, the wind and falling snow greatly increased. The return trip was made difficult by a complete white-out with the recent tracks completely erased by the blowing snow. I knew that the man and gear were close by, but I could not see them. I turned off the snowmobile motor and yelled. A muffled reply came back, but in the driving wind it was difficult to tell what direction it came from. We eventually made contact and, with the use of a compass, we worked our way back to the base camp through the whiteness.

In situations of this type, there are no tracks to follow back because the wind-driven snow can cover up all tracks in less than a minute. Remember that during a white-out, you should never let your group separate. Stay together because. once you get separated during the storm, it is very difficult, if not impossible, to get back together again. If you are traveling during these white-outs, you may have to improvise a snow shelter in or near a rock outcropping or a grove of trees, until the storm passes. Be very careful about traveling in these white-out situations. Because of poor visibility, it is possible to walk or drive into a creek, over a cliff, etc., and not even realize it until too late. It may sound exaggerated, but when skiing or snowshoeing during a white-out, I have seen visibility so poor that you could not see if the terrain was up or down until feeling it as you move forward. If not familiar with the area or without a compass, DO NOT travel during a white-out.

At night, during this type of a storm, be very alert if you are moving from one snowcave to another because it is easy to become disoriented in the dark. You might pass right by a snowcave if the entrance is covered with snow and you would soon be wondering around in a white darkness not knowing which way to go.

To show how easily this situation can develop, I recall a group tour on Park Creek. A blizzard moved in and lasted for several days. After the evening meal and discussions in the base shelter were over, the participants gradually retired for the night to their snowcaves a few yards away. After everyone had left the shelter, I turned off the lantern and prepared to go to my own snowcave a short distance away. As a force of habit, I first went to each snowcave to check and make sure that everyone was okay before I retired. Everything was pitch dark in the swirling snow of the storm because the people had crawled into their warm sleeping bags and blown out the candles. There was no light or glow from any of the snowcave entrances.

Somehow, while I was wallowing through the snow from cave to cave, I dropped my flashlight. It went out as it sank into the fluffy snow. I spent considerable time digging and poking around in the dark to locate my flashlight. When I finally did locate my flashlight and straightened up to move on to my own snowcave, I discovered that my sense of direction was gone. All of the surrounding tracks had been obliterated by the driving wind and snow. I could see no objects in the swirling snow to indicate direction and darkness left me with an eerie feeling. If I was to travel too far in any direction and not locate a snowcave, I would become hopelessly disoriented and might completely lose the campsite location in the darkness. I began to move a few yards in one direction and then immediately return to my starting point before the wind and snow covered my tracks. After a couple of attempts, I located a snowcave entrance and then was able to direct myself to my own snowcave.

This is a good example of why you should always attempt to remain calm and not let psychological stress take over and get you into more serious trouble. If I would have panicked and moved to far in the above situation, I would have soon become totally lost and then would have been in serious trouble. Even in good weather, always mark your snowcave entrance at night with an upright ski, snowshoe or a tree branch and handkerchief. Then if the entrance blows shut, you still know where it is.

The mention of blowing snow brings another word of advise to mind. When you are retiring to the snowcave or igloo for the night, always take a small shovel, flat stick, or some type of a digging tool inside the cave with you. Upon awakening the next morning, if you find the entrance hole blown full of snow, do not be alarmed. Air moves through fresh snow and you can still breath. After you are dressed and ready, calmly use the tool to dig out of the cave via the entrance hole.

This is another example of how snug a snowcave can be. You can go inside of an evening with the stars shining and poke your head outside in the morning to encounter a howling wind. Inside that snowcave, you do not even know it is storming outside. This is not the case in a tent where you might not sleep all night wondering when the wind will pull the tent stakes out of the snow and create problems as the tent flaps in the wind.

I would like to again emphasize another caution suggestion here, not with the intentions of spooking you away from winter outdoor enjoyment, but as a safety precaution. Always construct the snow- cave with a round dome which , like an archway, has a tendency to strengthen itself from above pressure. Do not leave the roof of your snowcave flat or it will start to sag. If there are old cracks or lines in the top of your cave that might collapse from a heavy new snow on top, move locations if it looks doubtful. If a new snowcave with a foot or two of snow in the ceiling would cave in, that is no great problem. But if a high wind or new storm deposits several feet of snow on top of a cave with a flat ceiling or other blemish, it could be a problem if all of that weight falls on a sleeping person. I have utilized over a hundred snowcaves and have never had one to cave in while sleeping in it, but that is not to say it cannot happen. Use common sense along with some judgement and learn to "read" the snow conditions.

It is recommended that you arrange your first few excursions with an experienced winter camper. That person can show you various technics that would take you several trips to learn. Remember that even when you are in the company of this seasoned winter camper, you should always maintain an open mind as you might come up with a better method. During this learning period of improvising snow shelters, you are not allowed as many mistakes in cold weather and blizzards as you are in the balmy weather encountered while summer camping. Use discretion and do not leave yourself or your group open for unforeseen situations with weather, terrain, and equipment. Your are your own destiny in the outdoors!

When on cross-country tours, start your snowcave construction before sundown

102

Group of snowcavers after a heavy snowstorm. Note two igloos in foreground

A correct camp location adds great comfort to winter camping. Cold air settles down and warmer air moves upward. So instead of making camp in the very bottom of a small valley or canyon, you might find a suitable location on a bench part way up the slope. The night air is warmer up there than at the bottom of the canyon. However, this is not always feasible because you might be exposing yourself or your group to a windy location instead of being where the air is still. During the summer, it might be of an advantage to set up camp on a point where the slight breezes help to keep the insects down. During the cold weather of winter camping, you do not need an added wind to carry body heat away. Occasionally, when you are choosing a snowcave location where the wind has created a large drift of snow, you will be exposed to some wind when you start to dig the snowcave. But once you get inside, there will be no wind.

Only experience can sharpen your skills on locating campsites, reading snow conditions, the type of equipment most suited for your uses, etc. Your comfort and enjoyment in the winter wonderland will increase accordingly as you gain experience. When you learn to be proficient with constructing snowcaves, you will in turn lighten your pack during winter excursions because you will not be carrying a tent. It is sometimes difficult to stake a tent down in the snow in order to withstand storms and winds. Another advantage of utilizing the snow for sleeping shelters is being able to sleep undisturbed. A snowcave is like a soundproof room and you cannot hear anything that goes on outside. In a tent, each new burst of wind or driving snow might make you wonder how long that tent shelter will withstand the elements. The temperature in a snowcave will remain constant all night long, usually just above freezing. By morning the temperature inside a tent will be almost as cold as it is outside.

Claustrophobia can be a problem for those who are new to using snowcaves. At night when you crawl into that sleeping bag everything is nice and cozy until you blow out the candle. Then it is very dark and very quiet. You cannot hear the blowing wind outside, and being confined in this kind of an atmosphere bothers some people. For some, a small source of light helps to lesson this problem. But I do not recommend leaving a candle burning while you are sleeping. It not only uses up some of the oxygen in the cave, but it could fall down onto your clothes or the sleeping bag during the night and start a fire.

I suggest that you occasionally crawl into the snowcave during the daylight hours and relax on top of your sleeping bag to adjust to the quietness. If the snow conditions will allow, construct the snowcave large enough to hold several bedrolls. Then the participants can lay in their sleeping bags relaxing while discussing the day's activities before getting drowsy and blowing out the candle. Your talking can continue in the dark until each one gradually drops off to sleep. This is a personal problem and is often difficult for an individual to conquer, but after my years of experimenting and becoming proficient at snowcave construction, I would not think of carrying a tent on my winter excursions, not even across the Continental Divide in Colorado.

After many years of winter camping experience, it generally takes me no longer to get a snowcave ready than it does other folks to set up a tent. The snow under that tent has to be shoveled out or tramped down level and then left to harden. It then takes a bit of effort and experience to make the tent stakes or anchors sturdy in the snow to withstand any wind against the tent. There are other advantages to utilizing a snowcave for a sleeping shelter. If your boots were wet the night before in a tent, they are probably frozen stiff when you attempt to put them on in the mornings. This is not the case in a properly constructed snowcave where the temperature will be the same in the morning as it was the preceding evening. Your boots or waterjug will usually not be frozen if they are set away from the entrance hole.

I have previously described several types of snow shelters, but before closing, I would like to again mention the small emergency snow shelter that could be useful during situations when the snow cover is shallow and time and energy expenditure is critical. Rather than taking the time and energy to construct a large pile of snow for a snowcave for several people, construct small individual holes in the snow. this would generally be used when alone and needing a sleeping shelter. Locate an upward slope that is not in an avalanche path. Then, you can lay an old log or some obstruction crosswise on this slope or tramp back and forth a couple of times in order to break the smooth surface of the snow. Next, from the upward side, scoot or toss the snow into a rectangular pile about eight or more feet long by approximately four feet wide. That log or rough depression left by the tramping stops the snow from sliding on down the slope. As you move the snow down to this pile, you might have to occasionally spread it out a bit wider rather than letting it come to a peak. If you make the pile about four to five feet high, that will be sufficient. Then after this pile of snow has been left for a few minutes to harden, you can start at one end and dig a hole lengthways as shown in the sketch. This hole should be long enough and wide enough to accommodate your sleeping bag.

Snow can be packed on the bottom of this long slanting hole to make it level for your sleeping bag. If you do not have a pad to go under your sleeping bag, put in a few boughs or pine needles and move your bed into the long hole. Close up the entrance hole with a backpack, boughs and snow or anything suitable, and crawl into your sleeping bag. You may not have the head space like you would in a typical snowcave, but you can manage if you are careful not to raise up too much and brush snow down your neck. You will not be on a raised sleeping bench, but you will find that shelter nearly as warm as a standard snowcave with less energy expanded during the construction.

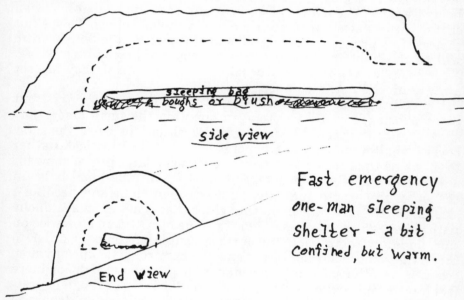

side view

End view

Fast emergency one-man sleeping shelter - a bit confined, but warm.

On a planned winter excursion with the time and tools for constructing more elaborate snowcaves, you can add assorted nooks to the inside of your cave if the snow conditions are suitable. Build an extra raised bench on one side for your packs and other gear or you might prefer to dig a hole at the side of the benches to slip your pack down into. By doing this the packs, boots, and other equipment are not falling down or crowding your sleeping bag at night, but they are still accessible when you need them. You can carve out small holes or shelves in the sidewall of the snowcave to hold your glasses, wallet, and other small items that you want to keep close track of. It takes little time and effort to form an extra hole in the snow of sufficient size for these different items. Again, use your own judgement and imagination for these added features, but as previously mentioned, do all of the digging or moving of the snow from these small nooks before laying out your sleeping bag so snow will not filter on it and make it damp.

As you can see, there is no end to the types of snow shelters you can utilize. When you are using one of nature's most insulating and available materials, your imagination and ingenuity is your only restriction. Get some practice or construction experience in your backyard, or a nearby drift up some side road before you completely depend on nature's bounty during an extended excursion into a remote area.

Learn to "read" the snow texture as it can change from day to day or from one area to another. If the snow is fluffy, tramp or pack the bottom layer before piling more snow on top to make a pile of snow for snowcave construction. Then that lower layer will not collapse, while digging out the cave, from the added weight of the added snow.

You might have seen my son, Larry, and myself constructing a snowcave on the TV program, American Trails, which has been shown nationwide several times. Seeing that program might have given you some basic instruction assistance to get started. Snowcave construction is not as easy as it appears to be and you will become more efficient with experience. It is not the Waldorf-Astoria, but it sure beats a cold tent flapping in the wind.

Try it! You will like it and more fully enjoy the uncrowded winter wonderland that awaits you when utilizing nature's blanket of snow.

INDEX

APPENDIX

Various items of clothing and other outdoor gear have been mentioned in the chapters of this book. If wanting some of these items and unable to locate them in your area, I have listed several mail order companies where special items can be purchased.

Early Winters, LTD.
110 Prefontaine Place South
Seattle, Wash. 98104

Recreational Equipment, Inc. (R.E.I.)
P.O. Box C-88125
Seattle, Wash. 98188

Campmor
P.O. Box 999-CC
Paramus, New Jersey 07652

The Coleman Company, Inc.
250 N. St. Francis
Wichita, Kansas

Marmot Mountain Works
331 S. 13th. Street.
Grand Junction, Colo. 81501

The North Face
999 Harrison Street
Berkeley, Calif. 94710

Wilderness Experience
20675 Nordhoff Street
Chatsworth, Calif. 91311

Damart Thermolactyl
1811 Woodburg Avenue
Portsmouth, New Hampshire 03835

Mountain Smith   (sleds)
12790 W. 6th. Place
Golden, Colo. 80401

Life-Link   (shovels, etc.)
Jackson Hole, Wyoming 83001

Yurika Foods Corporation
30700 Telegraph Road, Suite 2550
Birmingham, Michigan 48010

Chuck Wagon Foods
780 N. Clinton Ave.
Trenton, New Jersey 08638

Oregon Freeze Dry Foods, Inc.
P.O. Box 1048
Albany, Oregon 97321

American Footwear
One Oak Hill Road
Fitchburg, Mass. 01420

JanSport
Paine Field Industrial Park
Everett, Washington 98204

Names and addresses of numerous other outdoor equipment companies can be obtained from recent issues of outdoor type magazines such as "Backpacker" and "Outside". Then write for catalogs, compare, and make your selections.

These catalogs will also continually list new items and improvements on existing items.

The author conducts assorted summer and winter excursions in the mountains of Colorado for various school, college, family, and other groups. The summer trips are backpack tours into remote areas and include fishing, fire-building, wild edible plants, wildlife and tracking, and other related outdoor skills.

The winter trips include snowshoeing, skiing, constructing and utilizing assorted snow shelters, winter foods and clothing, wildlife knowledge and tracking, and numerous other related skills that can be included in educational and vacation excursions.

For full details and tour schedules, contact:

Ernest Wilkinson
3596 West Hyw. 160
Monte Vista, Colorado 81144
Phone 303-852-3277.